Natchez, Mississippi   AUBURN 1830

Stephen Duncan's house tells the story of ante-bellum life in the Deep South. Here today on a peaceful countryside, one still finds in these humble homes something of the grand manner, and in the grand homes—pure and simple elegance.

PLANTATION HOUSES AND MANSIONS OF THE OLD SOUTH

J. FRAZER SMITH

ILLUSTRATED BY THE AUTHOR
FOREWORD BY LEICESTER B. HOLLAND

DOVER PUBLICATIONS, INC.
New York

PUBLISHER'S NOTE, 1993

Although this book richly deserved to be reprinted for its architectural value, the present publisher deplores its occasional indulgence in racist reflections, whether these were conscious or otherwise.

Published in Canada by General Publishing Company, Ltd., 30 Lesmill Road, Don Mills, Toronto, Ontario.

Published in the United Kingdom by Constable and Company, Ltd., 3 The Lanchesters, 162–164 Fulham Palace Road, London W6 9ER.

Bibliographical Note

This Dover edition, first published in 1993, is an unabridged, retitled republication of the work originally published by William Helburn, Inc., New York City, in 1941 under the title *White Pillars: Early Life and Architecture of the Lower Mississippi Valley Country.*

Library of Congress Cataloging-in-Publication Data

Smith, J. Frazer (Joseph Frazer), 1887–1957.
    [White pillars]
    Plantation houses and mansions of the Old South / J. Frazer Smith ; illustrated by the author ; foreword by Leicester B. Holland. — Dover ed.
      p.   cm.
    Originally published: White pillars. New York : W. Helburn, 1941.
    Includes bibliographical references.
    ISBN 0-486-27848-4
    1. Plantations—Southern States. 2. Mansions—Southern States.
3. Architecture, Colonial—Southern States. I. Title.
NA7211.S63 1993
728.8′0976—dc20   93-25640
                                                CIP

Manufactured in the United States of America
Dover Publications, Inc., 31 East 2nd Street, Mineola, N.Y. 11501

In reviewing *the white-pillared houses of the Southerner, we are impressed by the apparent ease with which they housed his needs, and preserve for American culture his sincere appreciation of the art of living. Since they are all that remain tangible of his once potent civilization, it seems fitting that this book be dedicated to the parent Art*

Architecture.

FOREWORD

Whatever biologists may say to the contrary, the human race is to some extent a race of shell-fish. And as the expert conchologist can determine by studying an empty shell just what species of mollusc inhabited it, and how the internal anatomy of the dweller was organized and where and under what conditions it led its life, so it is generally easy by examining a human house to tell where it was built—if one did not happen to know—and what sort of man built it and under what conditions he lived. Generally, but alas, not always, for unlike the natural mollusc, shell-fish *homo* claims the privilege of free-will in designing his artificial shell. And sometimes the human oyster insists on trying to live romantically in a snail-shell, or vice-versa, or calls on his architect to concoct something wholly new, some cross perhaps between the conch-shell and the cockle. It is hard to say who, in the long run, suffers most in such cases, the architect, the client, or the community.

This book is a survey of the habitations of man of the Caucasian race—genus, North American; species, Deep Southern; variety, planter—thriving luxuriantly in the first half of the last century. His shells are characteristically American, related to those of his genus everywhere, in the northern colonies, in Virginia, Ohio, but developed to a special type by a special ecology, and flowering in peculiar splendor under particularly favorable conditions. No one else could have built these houses, nor could they have been built anywhere else. As neither the race nor the climate are essentially different today, it is not surprising to find many of these stately old mansions still comfortably inhabited. And many of the present generation, fashioning themselves shells adapted to new conditions, wisely follow the patterns which their pioneer forebears developed when they took root in these productive lands. How could they do otherwise? For the past as seen here is the templet for habitation, in the land of encircling verandahs, and will be until the live oaks shall vanish, or the race shall change.

Leicester B. Holland, F.A.I.A.

INTRODUCTION

THE history of America for the last three centuries is the story of pioneers discovering new lands and endeavoring to establish themselves on them. Generation after generation of Europeans, bold and ambitious, have pushed into untrodden regions of the North American continent, inland from the coast lines, westward from the Atlantic, northward from the Gulf, southward from the Great Lakes. Each wave of advance, breaking at its crest, has made a new frontier re-acting, with variations, the successes and failures of earlier ones. Different climate and economic problems modified the inheritance the colonists brought with them, and thus each frontier created a new design for living.

The most successful pioneers, the ones who became permanent settlers, have always been those who brought their women and children with them to establish homes in the new country. Our ever-advancing frontier has been a pageant of little groups. They traveled on horseback through the woodlands, or in flat-boats down the rivers, or in covered wagons over the plains. Each family carried its belongings to the new homes, kettles and feather-beds, treasured pieces of furniture and necessary farming implements. Building a house to live in, no matter how humble, gave immediate stability to the enterprise. Into the building of the home went the combined efforts of the whole group and, to defend it, men battled with strange environments and hostile natives, or even with other colonists.

The dwelling is the symbol of the character of our people. Not even our school buildings or churches have so continuously and so completely revealed our culture. Wherever a group of new comers to America has successfully adapted itself to an environment as, for example, that in Massachusetts, Virginia, or Louisiana, comfortable and beautiful houses have been built eventually. Wherever a group has failed to establish a satisfactory, permanent community life, as in the case of the early Spanish conquistadors, or the nineteenth century immigrants of our city slums, it has built no distinctive homes. A suitable style of domestic architecture is the hallmark of every successful section of the United States.

A colonial culture, that is, the way of life of any people who come from an old civilization into a new land, is naturally a complex matter. The newcomers bring with them an inheritance of ideas and habits. They are accustomed to certain ways of farming and tending cattle, to certain kinds of clothing and food and shelter. These habits they cling to at first, especially the more prosperous colonists who bring many possessions with them. But the new country forces them, or at any rate their children, into making changes. A bride of 1820 who took her trousseau made in Baltimore with her on a flatboat to her new home on a Cumberland River farm, found herself in a short time wearing homespun dresses and sunbonnets. A liking for codfish and baked beans as a diet had to be relinquished, even by ex-Bostonians, if they settled in a land where corn and hog meat and sorghum were more easily obtained. Orthodox church members, accustomed to frequent and formal worship, soon learned to get along with visits from the circuit rider and summer camp meetings.

No phase of life offers more striking evidence of this modification of old ways by a new environment than does domestic architecture. Indeed, it is the most apt medium through which to study the evolution of American culture, because it offers the most tangible data. To recognize the importance of the home as a social unit is to understand the part played by security, comfort, and prosperity in early American life. To view these early homes which have been preserved, and which show the various steps of their development, is to see the growth of a civilization. From the first hasty shelter erected almost overnight to the latest air-conditioned manor house of our own era, we in America have recorded our cultural evolution in our homes.

Naturally, each section of our continent has evolved its own kind of dwelling place. A steep roof to shed the snows of Northern New York is a useful feature, just as much so as is the wide gallery of a Mississippi plantation house for comfort during the ten warm months. Beauty as well as utility was served when these early homes were made to suit living conditions. A log cabin in the woods has its own simple form; a stout Cape Cod cottage is dignified and charming. Upstanding houses were rightly set against a background of tall trees in Alabama or Kentucky, whereas flat, one-story structures settled down more satisfactorily upon the wind-swept plains. Ways of building homes that "belonged" to their localities, functionally and aesthetically, were achieved by our forefathers chiefly as the result of their common sense, plus their inherited traditions and a fair knowledge of contemporary styles.

These early homes in the various parts of our country have been studied and emulated by modern builders with excellent results for our present-day architecture. Fine old homes, large and small, in New England, New York, Pennsylvania and Virginia have been accorded the respect which they deserve. On the whole, however, little has been done to recognize Southern architecture outside of Virginia. This is particularly true in the lower Mississippi Valley country which was the original Southwest, and is what we today call the Middle South and the Deep South—Tennessee, Kentucky, Alabama, Mississippi and Louisiana. Yet surely no section of our country pioneered more boldly or successfully in home building than this region. From the time of the American Revolution to 1861, these Southerners west of the mountains erected homes that confidently combined the old with the new, and united their traditions and learning with their desires and ambitions. Many of their houses still stand, despite the ravages of time and weather and war. In this book the reader will journey through a wide section of the South with an architect as his companion: from Lexington to Nashville, then along the Natchez Trace and down the great Mississippi River to New Orleans, with little by-way excursions to homes set back from the main highways. There will be little journeys, too, into the whys and wherefores of living in these old houses; for life in the far South, then as now, had a distinct pattern of its own. We will try to see that pattern as we go along.

The technicalities of architecture and its kindred arts and crafts will be discussed, and a final chapter will be given for those who care to delve further into the subject.

The illustrations (sketched in most instances on the ground) are sometimes unlike the present appearance of a house. An attempt has been made to place each house in its original and correct setting and to show it in its prime. When these changes have been made, the reader may be assured that the author has found sufficient evidence to warrant the change.

A house may be unlike the illustration in this book when you find it today; it may be under new ownership, restored, fallen to pieces, or perhaps disappeared altogether. If this be the case, you are asked to remember that this text has been in preparation for ten years. While an attempt has been made to bring everything up to date (fall of 1939), this has been possible only to a certain extent.

In spite of the rapid disappearance of these old homes over a period of seventy-five years, many thousands remain. It is impossible to illustrate and discuss, or even to mention, all of them. The examples discussed here have been chosen in an endeavor to present the particular life and its resultant architecture indigenous to each part of the several States.

Grateful acknowledgment is due Rebecca W. Smith, Professor of English, Texas Christian University, and Leicester B. Holland, Chief of Fine Arts Division, Library of Congress, for their invaluable contributions to this work.

To members of the office staff who have participated in various capacities: W. Jeter Eason, A.I.A.; Lillian S. Blair; William P. Cox; David McKinnie, Jr.; Robert T. Anthony; Irma Slaughter Brant.

To the many librarians and archivists who have given understanding assistance in historic research. And finally, to the owners of the houses discussed here, who have been disturbed in their peaceful occupancy by one or more of us.

To all these, the author acknowledges their contributions and expresses his sincere appreciation.

CONTENTS

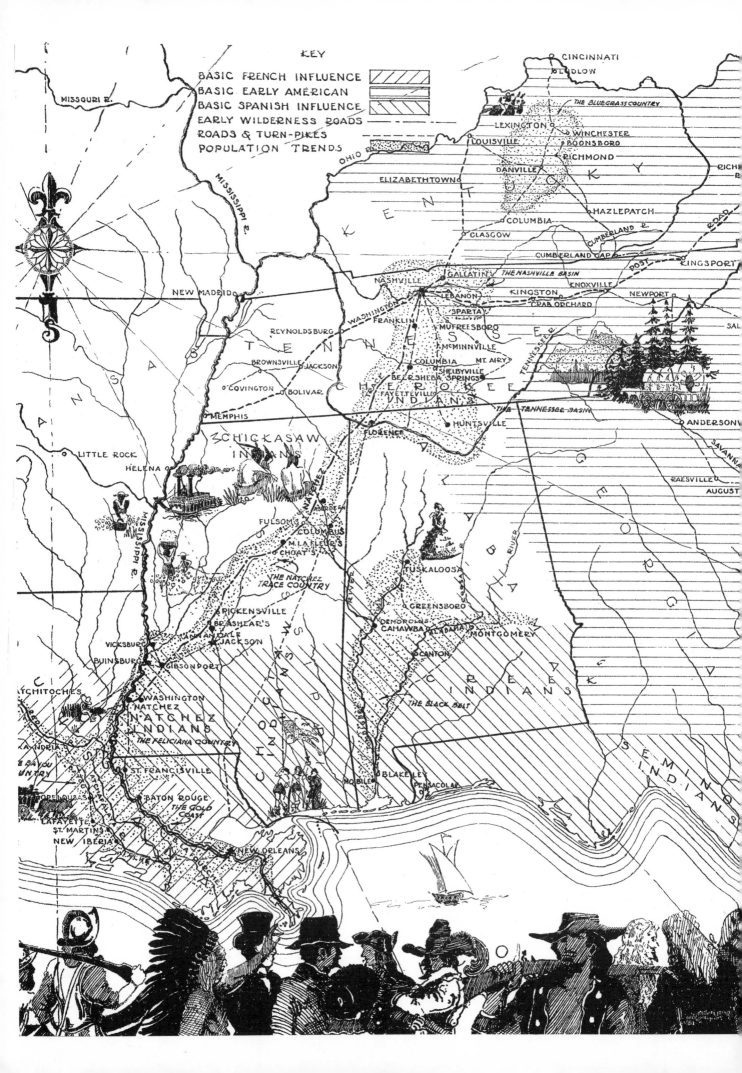

CHAPTER I

PATTERN OF LIFE IN THE OLD SOUTH

CHAPTER I

Pattern of Life in the Old South

IN 1850, an enthusiastic visitor from Massachusetts gave this vivid picture of the South and the Southerner:

"First, there is the slave himself, his condition, his cabin, his dress, his manners, his labors, his amusements, his religion, his domestic relations; then there is the plantation, with fences a mile apart, presenting in one broad enclosure land enough to make a score of Yankee pastures; then there is the cotton-plant, with its rich, pure, white fleecy treasures, hanging to the gathering hand; then there is the tobacco-plant, with its beautiful, tender green leaf in the spring, and its broad, palmetto-looking leaf in the autumn, green lined with brown; then there is the cotton-gin, with the negroes at work in it, the snowy cotton flying from the wind-fans in fleecy showers that mock a December snow-storm! Then there is the baling and screwing, the roping and marking with planter's name, all objects of interest to witness; then there is the planter himself, so different in his manners, tastes, education, prejudices, notions, bearing, feelings, and associations, from the New England man; then there is his lady, accustomed to have slaves attend upon the glance of her eye from childhood, commanding and directing her large domestic establishment, where the food, clothing, comfort, and health sometimes of a hundred slaves depend upon her managing care; then there is the son, who is raised half-hunter, half-rustic, with as much book learning as his pastimes in the field and wood will allow him to turn his attention to—the idol of the old negroes and the hope of the younger ones —who has never seen a city, but may one day walk Broadway, or Chestnut street, 'a fine young Southern blood,' with a fortune to spend, high-spirited, chivalrous, quick to resent an insult, too proud to give one, ready to fight for his lady-love or his country! prone to high living and horse-racing, but at home courteous and hospitable as becomes a true country gentleman; then there is the daughter of the house, too, a lovely girl, with beautiful hands, for she has never used them at harder work than tuning her harp, (and hardly at this, if she can trust her maid), who rides like Di Vernon, is not afraid of a gun, nor a pistol, is inclined to be indolent, loves to write letters, to read the late poets, is in love with Byron, sings Jenny Lind's songs with great taste and sweetness, has taken her diploma at the Columbia Institute, or some other conservatory of hot-house plants, knows enough French to guess at it when she comes across it in an English book, and of Italian to pronounce the names of her opera songs! she has ma's carriage at her command to go and come at her pleasure in the neighborhood, receives long forenoon visits from young gentlemen who come on horseback, flirts at evening promenades on the piazza with others, and is married at sixteen without being courted!"

Such a pattern of life would command and develop a suitable architecture.

Architecture must never be separated from its economic and geographic background. The land and the climate ultimately determine the kind of materials used, the height of the ceilings, and the cost of the building. If we are to understand the houses of the ante-bellum South, we must first have in mind a clear picture of the geography of the country. Along the Atlantic coast lies the tidewater region, narrow in Virginia but widening as it goes South, and including rich plantations of tobacco and rice. To the west of it rise the mountains, the Blue Ridge and the Alleghenies, which are not productive agriculturally, and have never supported their inhabitants in luxury. Still farther west stretch long, lean foothills across Tennessee and into northern Georgia and Alabama; in between these ranges lie three fertile limestone areas: the Bluegrass of Kentucky, the Nashville Basin, and the region along the Tennessee River near Muscle Shoals. Another fertile belt is the great crescent that loops below the mountains and hills, one tip touching South Carolina and the other stretching to the Mississippi River at Memphis. This belt is the heart of the cotton kingdom. And, too, there is the incredibly rich valley of the Mississippi all the way down to New Orleans, and the Black Belt of Alabama which extends from the Tombigbee to the Alabama Rivers. Cotton, tobacco, rice, and sugar—these were the money crops of the South, together with grain and cattle and horses that by 1800 were making the South rich, and by 1861 convinced her that she could hope to win political independence. Cotton more than any other one product poured wealth into the lap of the Middle South.

Other factors, too, conspired to exalt the agricultural South a century ago. Science contributed its share when the steamboat appeared to speed the traffic on the many rivers that flow across the region toward the sea. When Nicholas Roosevelt courageously captained the first steamboat over the Falls of the Ohio in 1811, and dared to descend to New Orleans in the teeth of a great earthquake and flood, the era of the steamboat began. For fifty years or more the white floating palaces, as they appeared to the wide-eyed inhabitants on the shores of the inland rivers, dominated the imaginations and lives of the Middle South. The cotton gin, invented by Eli Whitney in 1793, removed the last obstacle to the production of cotton on a huge scale. By 1840 railroads were clumsily beginning to transport heavy goods to market. No wonder men speculated in land and slaves when half a dozen consecutive lucky years would enable them to build magnificent homes near the town or along the rivers and to fill them with elegant furniture and china imported from Paris.

Whenever we think of the homes of the old South, we must keep the fertile regions in mind, for they will be the stopping places on our architectural journey. They were the localities where simple pioneer cabins were soon replaced by more permanent structures and often by mansions. Were we to make a map indicating the white-pillared houses of the ante-bellum South, we should be duplicating the economic map of the staple crops of the section in 1861. It is axiomatic that where there is wealth, architecture develops new forms rapidly. The domestic architecture of the old South, especially that of the Middle South, between 1800 and 1861, is a good example of the axiom.

Other geographic factors besides land must be borne in mind in discussing life in the South. The winters are, for the most part, mild or temperate; the summers are long and usually very hot. Rains are likely to be hard and beating; floods along the rivers occur intermittently and unexpectedly. Timber lands spread widely and vegetation grows easily. It is a section easily developed and often exploited and then forgotten. Such was the promised land into which pioneers came in great numbers for three-quarters of a century after 1776, a land of dramatic geographic and economic contrasts, where rich valleys and bottom lands are flanked by rocky hillsides and barren gulleys. Almost inevitably this would be a land of rich men and poor men, of great houses and of small cabins, side by side.

While the land and climate were powerful determinants of the architecture of the old South, so also were the people. Who were they? What traditions did they bring with them? Those who poured through Cumberland Gap and down the rivers into Kentucky, Tennessee, Alabama, and Northern Mississippi were mostly English and Scotch-Irish and Germans from Virginia and the Carolinas. They were pugnacious, ambitious, democratic. Andrew Jackson represents the type. On the other hand, the earliest comers at the mouth of the Great River and along the Gulf were French and Spanish by descent, with a language and culture of their own. Eventually the two streams of colonization met and blended along the Mississippi and in the bayou country of Louisiana. As a result, the domestic architecture of the Middle and Deep South is marked by a basic similarity underlying a multitude of surface variations. Because of a similar environment, the builders solved their fundamental problems in much the same ways; but because of different inheritances and minor variations in locale, the effects are delightfully individual and full of unexpected contrasts. A "Southern" house is unmistakable, even though it eludes precise definition, for the reason that it is functionally adapted to a particular way of life. Yet all classifications fail that try to define the types by arbitrary rules. Perhaps this is why textbook makers have neglected the architecture of the South—it is not easily pigeon-holed as Colonial or Greek Revival or Early Republican, and above all things it cannot be called "Southern Colonial."

A few generalizations may safely be made about the Southern way of life that is so large a factor in its home building. In a section that was made up principally of farms and small towns, land was plentiful to furnish yards, gardens, and even parks for homes. Family groups were generous in size and incessantly augmented by relatives and dependents. Houses were built on to whenever the need arose, or the masters' fancy dictated. All sorts of work was done on the place by and for the household, for which special buildings were provided; smokehouse, dairies, spinning houses, kitchens, carpenter shops, servants' quarters, and the like. Abundant slave labor made convenient and desirable the practice of having separate service buildings. Since the weather is mild, a house might sprawl at will with reckless disregard for heating facilities. Logs to cram the huge fireplaces were plentiful and so were stout darkies to chop and fetch them. Living was largely out of doors, especially in the Deep South; hence the numerous and spacious porches, windows, summer houses, and colonnades. The Southerner chose the site for his home either for its access to convenient transportation on road or river, or its elevation, from where he might see and be seen. Often he could accomplish both these ends by selecting a hill top overlooking the stream or highway. At any rate, his house, large or small, was pretty certain to have the stamp of his social and semi-public life, and to have provision for the coming and going of many persons.

The formative period of Southern domestic architecture, 1776-1830, was an era of great national pride and ambition. The winning of independence from Great Britain was followed by the erection of public buildings designed to be as impressive as possible. To Jefferson, Strickland, and other leaders in building, the style which best embodied the young nation's ideals was the classic temple. The states and counties often strove to emulate the federal style in building their capitols and courthouses. Therefore many an American who knew no formal rules of architecture vaguely associated the grandeur of the tall columns on his courthouse or his home with the glory of the political commonwealth of which he was a citizen. It fitted his conception of democracy that a successful man's house should resemble the Parthenon or, at least, the state capitol. Southerners were not unique in this era in admiring and appropriating the classic manner for their homes; the vogue for it appeared everywhere. However, they used it so independently and with such little regard for formal rules that they achieved more originality than did people of other sections.

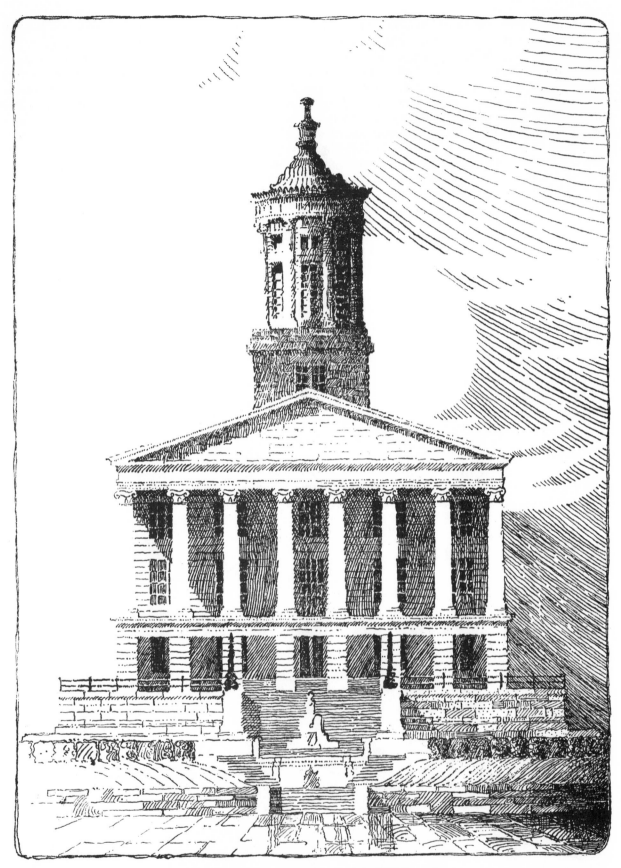

Nashville, Tennessee   TENNESSEE STATE CAPITOL 1845

This fine old building of Greek classical style shaped the destiny of white-pillared houses of Nashville and the Tennessee Basin, and extended its influence far into the Deep South. Visitors to the Tennessee Capitol building today are reminded of the epitaph on the tomb of Sir Christopher Wren, architect of St. Paul's, London: "Si monumentum requiris, circumspice" ("if you seek his monument, look about you"), for here too the architect, William Strickland, is entombed in the north portico of this, his last creation.

CHAPTER II

WEST OF THE ALLEGHENIES

PART 1. NASHVILLE AND ITS NEIGHBORS

TENNESSEE STATE HOUSE

A DOG TROT HOUSE   THE HERMITAGE

FAIRVIEW   BELLE MEADE

CRAGFONT   BEERSHEBA INN

BELMONT   MARYMONT

PART 2. KENTUCKY—BLUEGRASS REGION

DIAMOND POINT   WALNUT HALL

MOUNT BRILLIANT   ROSE HILL

CHAPTER II
West of the Alleghenies
PART ONE
NASHVILLE AND ITS NEIGHBORS

LANDS beyond the Allegheny and the Blue Ridge Mountains, known as the Southwest, lay south of the Ohio and extended to the Mississippi. Earliest pioneering efforts pointed toward two fertile and favorite regions. First, the Bluegrass region of Kentucky, which nestles close to the Ohio and extends southward, and second, the Nashville Basin, with Nashville as a radius point around which the Tennessee River describes a great semi-circle as it winds northwardly again towards East Tennessee.

These two regions grew more important throughout the pioneer period. A full and wholesome life developed around the centers which became a definite influence in the later development of the Deep South (the Gulf States).

Tennessee was a battleground for the social forces that shaped the old Southwest. Her domain swept from the little valley farms of the east through a productive limestone section to the low-lying bottoms and forests of the Mississippi River water-shed. In these different environments, her pioneer settlers developed strikingly different social patterns, each exhibiting those baffling combinations of democracy and aristocracy that are the key to Southern life, social and political.

Long before the Revolution, as early as 1740, restless dwellers on the Atlantic seaboard were yearning for the good land they had heard about beyond the mountains. Wealthy men sent surveyors to investigate possibilities for profitable speculations. Landless men everywhere—younger sons, poor farmers who did not own their acres, ambitious craftsmen, new immigrants—all turned their eyes west. Land was the measure of opportunity and power. A little of it was necessary; a lot of it was desirable. Scouts like Daniel Boone reported a marvelously rich domain to be had for the taking in the back country which vaguely was thought to belong to Virginia and the Carolinas. To be sure, there were obstacles such as King George III's express prohibition of all migration beyond the water-shed, and the peril of hostile Indians; but such hindrances were not enough to stop men who were land hungry. When they gathered up their wives and

children with their scanty goods and chattels to start across the ridges, they would not have paused, even if they had known what hardships lay ahead of them. The gaps through the mountains made rough going; there were innumerable streams to swim or ford; the Indians and wild beasts were equally terrifying. No matter, they went ahead. Before a shot was fired at Lexington, they were in what is now Tennessee and Kentucky, and at King's Mountain they contributed to the winning of independence. When the Revolutionary soldiers were paid with grants of western land, the tide of immigrants swelled rapidly.

In or about 1768, James Robertson, a stout North Carolinian of Scotch-Irish descent, and his little party lit their fires in a clearing on the banks of the Watauga River, tributary of the Holston, in what is now Washington County, Tennessee. Evan Shelby and John Sevier joined the Wataugans, and shared the battles of the group against Indians, and their struggles to establish a government, since neither Virginia nor North Carolina claimed Watauga Settlement. A decade or so later, Robertson likewise had a hand in establishing the Cumberland Settlements at the French Lick where Nashville now stands. He led a small party of men over Boone's Wilderness Trail through Cumberland Gap, and reached the Lick on Christmas Day, 1778, to find the river frozen over. The larger part of the expedition including the women and children were coming down the Tennessee under the guidance of John Donelson, a well-known Virginia surveyor, in the boat *Adventure*, accompanied by a flotilla of forty craft. Past dangerous Muscle Shoals to the mouth of the river and back up the Cumberland they fought their way. The trip was a hard one, especially for the wife of Ephraim Payton, who was delivered of a child en route, only to have it "killed in the hurry and confusion" of an Indian attack next day.

On Monday, April 24, 1779, Colonel Donelson recorded in his journal:

"This day we arrived at our journey's end and at the Big Salt Lick, where we have the pleasure of finding Captain Robertson and his company.

Nashville, Tennessee

THE CABIN AT BELLE MEADE

Double pioneer cabin, or dogtrot house, was an important step in evolution of the Tennessee and Kentucky white-pillared houses.

. . . Though our prospects at present are dreary, we have found a few log-cabins which have been built on a cedar bluff above the Lick by Captain Robertson and his company."

John Donelson made a clearing for himself at Clover Bottom about twelve miles from the bluff, where he planted a crop, which included the first cotton grown in Middle Tennessee. His daughter (the future Mrs. Andrew Jackson) would see her home, The Hermitage, stand upon a nearby site.

The cabin was the abode of the frontiersman. As we view today this log structure and its contents, we can reconstruct the life of the pioneer with striking approximation of the truth.

Thomas Perkins Abernethy in his *From Frontier to Plantation in Tennessee* gives us the following description of building a cabin:

"The cabin was, when possible, built by community effort. A common size for it was about twenty feet by sixteen feet, often with a low room or upper floor under the sloping roof. Work began by chopping down seventy or eighty of the tallest and straightest small trees in the immediate vicinity. When the felled trees had been chopped into proper lengths, the logs thus made were rolled to the site picked out. These preliminary processes required two or three days. Two logs, each sixteen feet long and of greater thickness than the others were then put in position twenty feet apart; at each end log a deep notch was cut on the upper surface extending through about one third of its diameter. Two other logs, each twenty feet long and correspondingly thick were next fitted with notches at the ends, both above and below, and were laid on the first pair, into which their lower notches were "dovetailed"*. A foundation was thus obtained that lifted the body of the cabin some three feet above the ground. About a dozen slender logs, sixteen feet long and usually ten inches in diameter, were laid at regular intervals so that they extended from one of the twenty foot logs to the other. These were to serve as a support for the thin slabs of wood that were later to be laid on them as a floor. The process with the large logs already described was then recommenced and each tier notched and fitted into the transverse timbers above and below until the walls had been built to a height about seven feet above the floor. Anothers row of slender logs was added at this point as the lap of the lower room and the floor of the one above. Three or four courses of the heavier trees completed the body of the structure.

* "dovetailing" was only one type; there were also the common "saddle-notched" and "square cut".

At either end of the upper framework a stout little tree, about six feet tall and so cut as to present two short diverging limbs at the top was set up, and from one such crotch to another the ridge pole was placed in position. The roof itself was composed of wide slabs of wood hewn bodily out of large trees and placed on the topmost tier of side logs with their upper ends converging and resting on the ridge pole. To keep the roof slabs in position, a long log was laid over the lower ends of each side of the cabin. Its extremities rested on the upper tier of end logs, which had been kept unusually long for this purpose; and it was in turn held secure by heavy wooden pins. Other timbers over the roof slabs were placed in similar manner, and the body of the cabin was complete.

The doors and windows were sawed out after all logs were in place, and their edges were cased with slabs to keep the walls from sagging. There was no glass, and all openings were protected by strong doors. The window panes were made of paper, when it could be obtained, plentifully coated with hog lard or bear grease. The big fireplace was constructed of large, flat stones, and the chimney was built of sticks laid in the same alternating manner as were the timbers of the house, with the chinks of the chimney strucsture filled and covered with clay that was soon hardened by the heat. All spaces between the logs were then stopped up with mud and moss and generally plastered over with clay in addition. Slabs were laid for the floors; a perpendicular ladder of five or six rounds served as a stair-case; and the domicile was finished.

The whole job took about a week, but was often done in less time if six or eight men were working. Not a scrap of metal had entered into its construction. It was wholly a product made from materials found within a quarter of a mile of the spot where it stood ready for occupancy."

As the pioneer became a permanent, self-sustaining settler, he advanced from the one-room log cabin to the "dog-trot," or "possum-run" house. In plan this was a two room cabin with a run or open hall between. Later on a front porch and rear porch were added, but always the two rooms were the basic plan. While the early cabin dweller slept, cooked and ate in one room, as soon as times became a little better, be wished to observe the proprieties and to divide the living quarters of the sexes in his large family. Then, too, when relatives and friends came visiting, they necessarily came long distances and had to be housed. On such occasions, the men occupied one side of the "dog-trot" house and the women and children the other, while the dogs belonging to the household and to the visitors occupied the run between, and barked their warnings if any disturbances arose outside. During mild weather the family often ate in the run. If much company came, it was used by the young folk for dancing. So close was the wilderness that, in the absence of the dogs, an occasional opossum would prowl through the run in search of food; hence the common name "possum-run" for the passageway.

John Sevier, the heroic pioneer of East Tennessee, has left us a Pepysian record of the life of a prosperous, busy family on the frontier. He wrote in his Journal during December, 1795, these items:

"Sun. 6—cold & clear in the day but stormy and began to rain towards day, sent Jim to Jonesbo for R. Camples negroes. Mon. 7—re-

cloudy in morng. Mon. 14—Some snow in morng. began to kill Hoggs. Tues. 15—cold, Killed Hoggs 16 in 2 days. John Fickee to 1 pr. stockngs got in Harrisons store Price 16. Finished walling & Plastering the Cellar of the Kitchen. Wm. 200 ls Flour of Wm. Clarke at 12 pr. ct. Wed. 16—James laid the kitchen flour. Mrs. Sevier & R. Campble wt to Jonesbo. Thurs 17—I killed a large turkey cocke. cloudy. Fry. 18—went to the Election. Sat. 19—tarried at Jonesbo. Let John Keele have 2 dollars."

TYPICAL CABIN PLAN

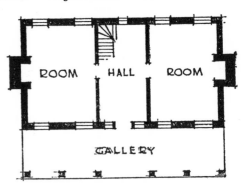

"GLORIFIED PIONEER HOUSE"
FIRST FLOOR PLAN

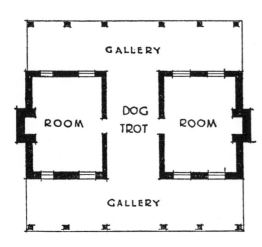

TYPICAL DOG·TROT HOUSE

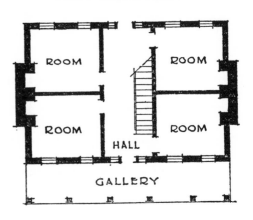

TYPE OF FINAL DEVELOPMENT
FIRST FLOOR PLAN

markable high winds with some rain. Josiah Allen began the kitchen Cellar. Tues. 8—more moderate. Wed. 9—Mr. Debardelabins family arrived, & took their Horse and negro boy away and Got 2 bushels of corn & half bushel of meal. Mrs. Davis, wife of Nathanl. Davis died & is to be buried on the 11th inst. Self and Mrs. Sevier Dined at Mr. Sherrils. Thurs. 10—I went to Jas. Seviers to Hunt turkys. R. Campble Rutha & Washgn went to Jonesbo. Fry. 11—cold morning & hard Frost. James Anderson came here in the evening & tarried at night. Sat. 13—windy Washington R. Campble & Js Anderson went to Jonesbo & Returned in the evening & tarried all night. Rain in evening & all night. Sun. 13—

It is easy to see why the Seviers found it necessary to busy themselves procuring corn and hog meat as well as building a kitchen, for this fortnight records only a usual quota of their guests who "tarried all night." What does surprise us, however, is the amount of traveling that they found time to indulge in. Life in a well-to-do pioneer log house was busy and far from dull.

From the start, some pioneers prospered more than others, and these promptly sought to provide extra comforts for their families, especially for the women folks. Such a prosperous settler, still primitive in his methods of building and still with only native materials at hand, developed his house plan for convenience and comparative luxury. A kitchen, a dining room, and even a parlor appeared in due time to enrich the social, religious and certainly the matrimonial prospects of the family. However, the typical "dog-trot" plan was never discarded; it was simply enlarged by adding another story. The house was still built of hewn logs fitted together after the fashion of a log cabin. On the ground floor two rooms faced the front with a wide (first open and later enclosed) hallway between, which housed the stairway. The plan of the second floor was the same as that of the first. A porch of medium width spanned the front of the house. This porch might be either one or two stories high, and was usually covered with a slanting roof. A single gabled roof crowned the main structure, and the hand-split shingles were fastened in place with wooden pegs. There was a stone chimney at each end of the house, and no windows pierced the walls at these ends. All openings were to the front and rear. A kitchen was often appended as an L to the rear, and as the family grew the house grew, other rooms being similarly added. Plain planking was used for floor and woodwork, and as materials and tools became more accessible, framework and planking or weather-board might be substituted for the log construction. Natural stone and brick appeared here and there, but the plan remained the same.

Prosperity came to Nashville, originally called Nash's Station or Nashborough, because it commanded the navigation and trade of the Cumberland River, but it continued to be a frontier post for many years. In 1785, Lewis Brantz, a seventeen-year-old German traveler, wrote:

"Nashville is a recently founded place and contains only 2 houses, which, in true, merit that name; the rest are only huts that formerly served as a sort of fort against Indian attacks."

In 1791, a brick house had been built, and in 1800, the great apostle of Methodism, Bishop Francis Asbury, found a thriving community. In his Journal he noted:

"Sunday 19. I rode to Nashville, long heard of but never seen by me until now; . . . not less than one thousand people were in and out of the stone church; which if floored, ceiled and glazed, could be a grand house."

All travelers to Tennessee visited Nashville, and inn-keepers did a thriving business. The eminent French botanist, Andre Michaux, was apparently at the mercy of his landlord in 1796, when he jotted down in his "Travels" what his stay cost him:

"Prices at Nashville; Dinner 2 shillings, Breakfast or supper 1 shilling 4 pence; ½ Quart of Whiskey 1 shilling. Hay and maize for Horse 2 shillings. The whole of 6 shillings for 1 Dollar."

The Duke of Orleans, later to rule France as Louis Philippe, when he visited the town during court week in the spring of 1797, found that one bed had to do for three. There was some solace, however, in the fact that it was possible to get good whiskey to fill his canteen which he wore around the princely neck, while traveling thence to Louisville.

By the turn of the century, cotton and tobacco led the exports, grain- and stock-raising were on the increase, and good times had arrived for the larger land owners, merchants, and professional men. Entertaining on a lavish scale flourished from the very early days, colored often enough with a political significance. James Robertson at the start had built his home in the country, and leading landed gentry established their places on the many good limestone roads that radiated in all directions from the town. Judge McNairy, on the other hand, maintained a handsome town house as well as a plantation domicile. General James Winchester's Cragfont on the road to Bledsoe's Lick was admired by all visitors, as was General Daniel Smith's Rock Castle, east of the Winchester holdings on the same road. These houses, to the eye of the architect, mark a transition from the pioneer cabin to the glorified pioneer house, a house of late Georgian influence built along native lines. From the viewpoint of the social historian, they indicate the trend of life in the rich Nashville area to shape itself into an aristocratic pattern similar to that of Tidewater Virginia and Carolina. East Tennessee remained, until long after Appomattox, a land of small farming gentry and poor hill folk; but Middle Tennessee was on its way to riches, as its export crops floated down the rivers to the Gulf and the world beyond.

As we visit houses in and around Nashville and in the Bluegrass country, it is interesting, architecturally, to observe how fully this development of the Tennessee Basin plan and façade can be detected: this evolution from cabin, through "dog-trot" house, through glorified pioneer house, to aristocratic plantation mansion.

As to plan: wings may project to the rear in forming service rooms; to the side, or, to balance, one on each side. Various additions served greater living needs, but the basic characteristics are always available in the hub or central motif.

In elevation, Tennessee and Kentucky white-pillared houses are unmistakably characteristic in their closed gables, pediment-porticoes and outside chimneys.

Near Gallatin, Tennessee   CRAGFONT 1802

This is Tennessee's glorified pioneer house. General James Winchester remembered the late
Georgian Period in New England.

CRAGFONT

Cragfont, a typical glorified pioneer dwelling executed in native stone by General James Winchester, reveals this transition in social life as completely as it does the character of its distinguished builder.

James Winchester, born in 1752 in Carroll County, Maryland, of an aristocratic family of English descent, saw vigorous service during the Revolution under Washington and Greene. Tradition has it that

Winchester was prominent in many phases of frontier life. He held commissions in the militia, he operated a large mercantile business with connections down the rivers, in the conduct of which he fostered the early steamboats on the Cumberland. It is believed, too, that Aaron Burr came to his home in a futile attempt to interest him in dreams of empire.

It was apparently in 1802-03 that Winchester

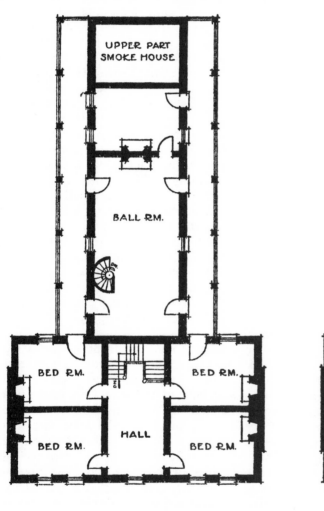

SECOND FLOOR

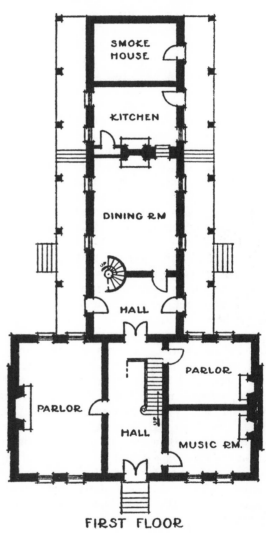

FIRST FLOOR

CRAGFONT

it was from fellow soldiers in the army in the South that he heard tales of the rich Tennessee lands. At all events, he arrived in Nashville in 1785 and chose a fine site on a creek a mile from Bledsoe's Station, in the midst of a native forest full of game, and crossed by Indian trails. Within three years he built a large mill, married a charming young lady of the county, and established a home on the site now occupied by Cragfont.

erected the magnificent stone residence, Cragfont, which still stands in good condition. In September, 1802, F. A. Michaux, son of the famous traveler, André Michaux, passed by the place and, as a result, wrote the following interesting description:

"We likewise saw, *en passant*, General Winchester, who was at a stone house that was building for him on the road; this mansion considering the country, bore the external marks of grandeur; it consisted of four large rooms on the ground floor, one story and a garret. The workmen employed to finish the inside came from Baltimore, a distance of nearly 700 miles. The stones are of a chalky nature; there are no others in all that part of Tennessee except round flints. . . . On the other hand, there are so few of the inhabitants that built in this manner, masons being still scarcer than carpenters and joiners."

At the outbreak of the War of 1812, General Winchester shared with General William Henry Harrison the command of the American troops in the Northwest; but he won none of the military glory that sent "Old Tippecanoe" to the White House. Nevertheless, when Winchester came back to Nashville, he was still esteemed and defended by his fellow Tennesseans. He helped to found Memphis, and was active in state affairs until his death in 1826 at his beloved Cragfont.

On his tomb in the garden there was inscribed this rhetorical but sincere tribute from one of his friends.

How oft, alas! we see the worthless name
   Bedecked by fraud with trophies of the brave;
While lost, forgotten, or unknown to fame,
   Oblivion's wing obscures the patriot's grave.
. . . .
But when false claims to glory meet their doom,
   And Truth, with clarion note, once more shall rise,
The historic muse will point to this lone tomb,
   Where native worth and spotless honor lies.

A visit to Cragfont is equivalent to reading a page from cultural and architectural history, for this fine old pioneer house illustrates the evolution of a natural plan of building on the frontier. Here a man came from the old world of custom and refinement, adjusted himself to keep step with a backwoods settlement, and found that he could build his home in no better way than to modify the pioneer plan which had been developed by local living requirements and experience. The long drive up a dangerous and rocky lane from the highway to Cragfont is a fitting introduction to what is to come at the end of the drive. Situated on the very top of the craggy hill above Bledsoe's Creek, the house looks down on the approaching visitor with all the hauteur and dignity befitting the position of eminence occupied by its builder.

It is a large house presenting a bold gray stone façade after the Georgian manner. The stones are laid evenly and closely, broken only by the necessary window and door openings. Very little detail is employed and that little is in excellent taste and well executed. The entrance is accentuated only by a simple transom and stone steps leading up to the door. There is a large stone chimney at each gabled end, so wide that each embraces fireplaces in two separated rooms. No windows pierce the gable ends. Extending to the rear of the house is a long two-story wing which forms a "T." Two-story porches on either side of the wing, but not extending around the end, terminate against the front section of the house.

On the interior, a typical center stair hall is flanked on the left by a large parlor extending across the entire end of the house, and on the right, adjoining the music room, by another parlor. The parlor to the left is entered through a large doorway with panelled jambs, opposite which is a great mantel extending from floor to ceiling and elaborately executed in delicate woodwork. A rich chair rail extends around the room at the height of the window sill and is a continuation of the same mouldings. Great care was exercised in this feature, as the sill and apron are of marble, matching perfectly with the rail. The windows are set in deep panelled reveals, out of which open the interior shutters in sections divided at the meeting rail of the sash. The same details and careful workmanship are carried out in the other rooms of the first floor.

After leaving the front, or main section, we find that Cragfont departs from the conventional Tennessee pioneer plan. The reader will observe that at once by examining the plan drawings of the interior. At the rear of the central stair hall a door leads to the cross hall of the wing. This hall affords circulation between the two porches on either side of the wing. Immediately beyond this cross hall is the dining room; back of it, and slightly lower, is the kitchen. Not connected by a doorway, but separated by a wall is the last room, the smokehouse, which extends the full two stories. The location of the smokehouse in the residence proper is (so far as we know) unique.

The second floor of the front section is typical of the period and, except for unusual care and workmanship, differs little from other houses of the time. The second floor of the wing is not accessible from the second floor of the front section, but is reached by means of a small, circular stair located on the first floor between the cross hall and the dining room. This circular stair leads into the famous ballroom, which occupies all the space above the cross hall and dining room. General Winchester's gracious hospitality was expressed in the building of this enormous ballroom, probably the first in a private home in the state. There is a remarkable resemblance in the plan location of this ballroom to that of the Governor's Palace in Williamsburg. Members of the old families in the Nashville section still recount stories of ladies and gentlemen dancing out through the doors of the ballroom on to the porches to seek romance in the soft light of a summer moon, while the strains of violins floated over the fragrant gardens. The ballroom opens on to the second floor porches on either side, affording an ideal view of the gardens and lake in the grounds and of the farm lands in the distance.

Leaving the house, we find fragmentary remains

and indications of the once luxuriant gardens that lay between the lake to the back and the almost sheer slope in front. Letters, stories, and these few physical facts show that the garden was to the right of the house and was reached by a main walk laid at right angles to the rear wing. This walk apparently led to a tea-house and was intersected by a maze of walks leading to rose gardens, beds of strawberries and raspberries, and terraces. The gardens of Cragfont were reputed to be the most pretentious in the state at that period, and did credit to the General's ambition to be master of a home and estate comparable to those he had known in Maryland. The grounds around the house were graced by large trees, and a center walk to the door was lined with cedars. To the rear and left was the family cemetery. The auxiliary farmhouses, the barn, the springhouse, the servants' quarters and the office were of stone.

Other homes of the period in the neighborhood of Cragfont are of similar plan and construction. Rock Castle, built by General Daniel Smith in 1784 on land granted to him by the State of North Carolina for service in the Revolutionary War, is of stone also, and has the marks of having been constructed by the same masons. Spencer's Choice, near Gallatin, was built by Colonel David Shelby on the grant of Thomas Spencer, one of the first settlers in the Cumberland Country in 1777. This house, too, is of stone and was built about the same time as Cragfont. Station Camp, home of the Peytons, was constructed of brick by John Peyton about the time of the erection of Spencer's Choice, with virtually the same plan and elevation. Colonel Baylie Peyton added a white-pillared veranda in 1840, and it became a famous hostelry in that prosperous period. It was demolished some years ago.

Such were the homes of the last of the pioneers of Middle Tennessee, built at the beginning of the nineteenth century. They were typical of the period, as sturdily indigenous as the virgin timber and native stone that went into their construction. These homesteads portray frontier society at its height, with land already becoming concentrated into large holdings, and marriages among the better families carefully planned. Into such houses as these a newcomer like Andrew Jackson was welcomed; and in their wide halls he came quickly to realize that his immediate goal was to become master of such a place. The boy from the Waxhaws fixed his ambitions upon a Middle Tennessee homestead long before he though of being President of the United States.

THE HERMITAGE

THE courthouse at Nashville was a log cabin eighteen feet square on that October Sunday in 1788, when twenty-one year old Andy Jackson from North Carolina rode in town to become Judge John McNairy's public prosecutor, and to put the fear of the law into the debtors of the community. He resided at the loghouse home of the Widow Donelson six or seven miles across the river, not only because of the good lady's noteworthy table, but because of her beautiful daughter Rachel, lately separated from her husband, Dr. Robards. How Jackson and Robards quarreled, how Rachel fled to friends in Natchez, how Jackson wooed and married her there after her husband divorced her—all that is part of Jackson's dramatic career. The bride and groom, each twenty-four, honeymooned in a log house at Bayou Pierre, overlooking the Mississippi near Natchez; then Jackson brought Rachel back home to Nashville.

At once he bought from John Donelson the Poplar Grove Plantation, at Jones Bend on the Cumberland, with its loghouse residence. Before long, having prospered mightily as judge and merchant, he built a fine house on his other plantation at Hunter's Hill, a real step toward the Tidewater manner of life. Extravagant new furniture was bought by the judge on a trip to Philadelphia in 1804, but financial reverses came before Rachel could unload her parlor chairs and settee. Hunter's Hill was sold. The Jacksons moved to a six-hundred and forty acre tract called the Hermitage that he had bought in 1795, where Rachel once more lived in a loghouse, remodelled but much less fine than the Hunter's Hill home. Her harpsichord doubtless looked odd in the one, big down-stairs room, dingy and roughhewn; but we are told she was even happier here than at Hunter's Hill.

In the loghouse on the Hermitage estate they lived glad, exciting years until 1818 when the General, by now the hero of New Orleans and the Indian Wars, determined to build a proper mansion on the Hermitage property. "Mrs. Jackson chose this spot," General Jackson is quoted as saying by Marquis James. "The house was built in the secluded meadow." One feels sure that the General would have selected a more commanding spot, but her wishes were sacred to him.

This 1818 Hermitage was a fine house, as befitted the chief citizen of Tennessee, through whose doors came most of the great men of his time. Downstairs were parlor, dining room, and the family chambers;

THE HERMITAGE 1831

Near Nashville, Tennessee

Andrew Jackson's dream of a white-pillared house is said to have long preceded his dream of the Presidency.

above were four great guest rooms that were rarely empty. The service quarters were all separate at the rear. In the spring of 1819, an English gardener laid out the grounds in rolling lawn dotted with great trees, with a guitar-shaped driveway lined by cedars leading to the porch, and a large flower garden. The boy from the Waxhaws had achieved the grand manner of living.

Even this new home, however, was not so fine as Jackson wished. While he was President, he caused the house to be remodelled in 1831, by adding a one-story wing on each side and a large porch across the front, thus making it larger and more adaptable for entertaining on a lavish scale. The house was partially destroyed by fire in 1834, but the foundations and walls were usable, and the present Hermitage was built from those foundations. It now stands as it was at the end of Jackson's career. The log cabins on the place remain as they were during pioneer days.

Thus one man, Andrew Jackson of Tennessee, lived and built through both the pioneer and plantation periods, from loghouses to white pillars. The house, garden, tomb, service houses and cabins tell the story of his life. As the old man lay dying at sunset of a June day in 1845, surrounded by his slaves and kinsmen in the Hermitage, he may well have had memories of his mother spinning in a log cabin in the Carolinas, of Rachel in the old log-house playing with little Andrew, of Rachel protecting the itinerant artist Earle for seventeen years while he painted portraits of the Middle Tennessee gentlefolk, of his own lordly days in the White House . . . and always as the crown of his achievement, his position as master of the Hermitage, the most famous mansion in Middle Tennessee. Not only to the student of Jackson's life, but to

the architect as well, a visit to the Hermitage is noteworthy. From the approach, it presents a delightful silhouette as it grows from a one-story wing, flanking each side, into a two-story colonnade and parapeted main portion. The Hermitage seems to have set the pace as to silhouette in the Tennessee country, for later we find this harmony of façade in Sunnyside, Belmont, Belair, the Caruthers place, and the Nathan Green homestead in the Nashville vicinity. It is also noticeable in Beechlawn, Merca Hall, and many others along the Natchez Trace.

The order applied at the Hermitage, on the south façade, is modified Corinthian. The columns are of fair Roman classical proportions, fluted and correct of cap and base. The somewhat simplified entablature is pleasing in proportion. The colonnade as a whole is acceptable until we move around to the side elevation and receive a let down. The ensemble indicates a flat roof; instead, there is a very definite pitched roof, terminating in a gable on the opposite north elevation. In this elevation the pitched roof is correctly expressed in a lovely pediment supported by six fluted Roman Doric columns, each on its own pedestal and in very good proportion, spaced at uneven intervals. The south portico, however, denies the existence of a pitched roof and, accordingly, is automatically subjected to a very old rule of architecture: "Architecture to be good must be honest." The façades must agree and all must express the true plan. I find the two one-story portions very pleasing. Their simplicity properly glorifies the important orders of the portico, and their one large opening is well placed.

The fenestration is simple and in good taste. Spacious double entrance doors with side-lights are framed simply and in keeping with the main chosen order. The full second-floor balcony becomes a veranda, supported by the columns and bordered with simple square balusters. The whole veranda is

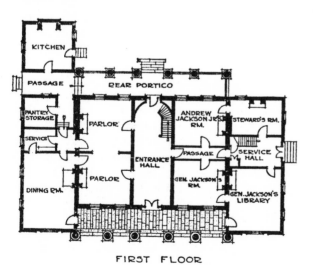

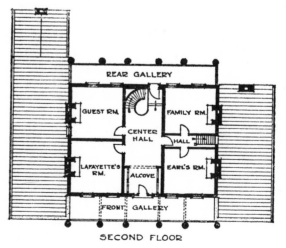

FIRST FLOOR

SECOND FLOOR

"THE HERMITAGE"

white with the exception of the blinds, and is very effectively set in its natural green entourage. The materials of wall construction are brick with a stone foundation. The columns are of wood, as are the cornice, trim, and porch work. The roof, too flat for shingles, is wisely covered with tin.

We enter, by way of the veranda from the front, a spacious stair hall. At the opposite end a free-standing, wood stairway leads to the second floor in one grand sweep, and seems to form a S-curve from the point at which it is viewed. Simple, wooden balusters with a one-piece handrail of mahogany add grace to the curves. The doorways with rich white-panelled jambs are otherwise simple and are hardly noticed because one immediately becomes absorbed in the theme of the wallpaper, which depicts the search of Telemachus for Ulysses. The plank flooring is painted dark. The ceiling has a light plain surface, conspicuous for its absence of decoration and cornice, an aid in holding interest in the stairway and walls.

The hall on the left opens into the front and back parlors, which in turn lead to the dining room. The parlors are graced by crystal chandeliers and Italian marble mantels with conventional mirrors. As we stand here, we have little difficulty imagining Rachel Jackson's interest in selecting furniture and house-hold luxuries in New Orleans in 1821 for her growing household. Seven cases of goods, totalling a freight bill of two hundred and seventy-three dollars and seventy-five cents were unloaded in the house—silver and mahogany, soft mattresses and fine brandy.

In the dining room is furniture of Early American and English influence. The celebrated "Eighth of January" mantel-piece is a rustic affair built of pieces of hickory by one of Jackson's veterans of the Battle of New Orleans. Since this was purely a work of sentiment, the craftsman did all the work himself, working only on the days of the anniversary, until he completed it on January 8, 1839. General Jackson installed it on January 8, 1840.

The Old General's bedroom is the most personal room in the house. It awakens a certain tender respect for it possesses much that is human. There is a variety of Early American furniture. The massive old four-post bed, with little steps at the side, is topped with a rich fringed canopy to match the draperies at the windows. A favorite couch by the window is where he spent much of his time during the late years of his life. His chair, his dressing gown, and his tobacco box are all so personal. He placed the portrait of Rachel over the mantel where he could view it the first thing in the morning and the last thing at night. It was on this picture his closing eyes rested the day he died.

The combination office and library adjoining adjoining General Jackson's bedroom was the center of political activity for many years. This is a livable room with a mixture of furniture consisting of a Spanish Renaissance chair, Duncan Phyfe table, two Sheraton desks, a table, and a comfortable chair, known in those days as an "invalid's chair."

The bedrooms upstairs are all richly furnished in the same elegant tone as the master's bedroom.

The wall-to-wall carpets are in character with the period, rich in the spirit of the past, with their lavish color and design.

The plan of the Hermitage displays a complete establishment which evidently functioned satisfactorily in every phase of the family and public life demanded by such a famous personage as the General. Noticeable is the service wing which, as in so many houses in Tennessee, is attached to the house by a passage instead of being completely detached. The smokehouse, in all cases a most important accessory to a Southern plan, is near the house. The servants' and carriage houses complete the immediate group.

The gardens, at the right of the entrance and in full view of the library, while not so conspicuous as those of some of the nearby places, are a restful spot. Mrs. Jackson loved flowers, and exchanged plants and seeds with her friends and neighbors. Said to have been designed and laid out by Ralph Earle, the artist whom the Jacksons befriended for years, this garden plan is a Greek cross of brick walks lined with beds. In the spring there are lilacs, bridal wreath, calycanthus, purple Japanese magnolia, fringe tree shrubs, jonquils, narcissi, hyacinths, lilies-of-the-valley and tulips. In the early summer come honeysuckle and roses of many colors and varieties common to the South, literally covering the fences, arbors, and trellises in every direction. Pink crepe myrtle grows profusely everywhere and suffulicosa box-plants add to the borders. In a quiet, far corner is the tomb erected by the General for Rachel Jackson. It is one of the most beautiful spots in the garden, its white Grecian shafts flanked by hundreds of colors of the flower she loved so much.

On the left of the entrance to the garden, under a weeping willow, is a bronze slab embedded in stone, which bears a message from President Andrew Jackson to Sarah York Jackson (Mrs. Andrew Jackson, Jr.), Hermitage Mansion. It is as follows:

"White House"
May 19, 1832.

"I sincerely regret the ravages made by the frost in the garden and particularly that the willow at the gate is destroyed. This I wish you to replace. The willows around the tomb I hope are living, and a branch from one of these might replace the dead one at the garden gate. It will grow if well watered and planted on receipt of this."

Following this is the emblem of the Daughters of the American Revolution and this statement: "This willow planted by the General James Robertson Chapter, Daughters of the American Revolution—1926."

BEERSHEBA INN

Frontier folk were habitually on the move. They made trips back to the older sections for supplies; they visited other settlements; they went on political and cultural errands. Middle Tennesseans, in the early days, were no exceptions to this rule, nor were they slow to develop means of transportation. The rivers were dotted with keels, flatboats, and later, steamboats, establishing far-flung connections with New Orleans and Europe. The inland roads opened up new territory. Indian trails were widened into packhorse roads and by 1785 were improved for carts and wagons to cross the mountains.

About 1820, the turnpike and plank road appeared in Tennessee, and ten years later the first local stages were operated. In 1839 the Gallatin Turnpike inaugurated a new era of travel in the region, for it eventually extended forty-nine and one half miles to the state line, connecting with the Louisville Pike and so permitting daily stages between Louisville and Nashville. This grand-scale road construction cost $290,000, and became a cosmopolitan artery which brought the North and South into a daily contact.

The coach was pulled by four horses, always in a gallop and exchanged every ten or twelve miles for fresh ones, which were harnessed and held in waiting by four quick groomsmen. Twelve passengers could ride inside and five on top. Baggage was carried in a "boot." From Louisville to Nashville took two days and a night, and the fare was twelve dollars. The stage driver of course became a hero to the countryside. Indeed, the turnpikes were busy and colorful. Private conveyances with their regular cavalcades of servants, outriders, and baggage wagons journeyed from plantation to plantation, or city, or resort, lending to the highway a touch of gentility. Big freight wagons with sky-blue beds carried thousands of pounds at a load under the direction of two drivers, who carried with them provisions for the whole trip, which sometimes required days or weeks. Four miles an hour during the day was their speed, and they camped out at night.

With the development of highways about the year 1830, inns and taverns sprang up over the South. Some were commercial along regular stage routes, and some were safely tucked away off the main lines leading to resorts at watering places and scenic points, catering to the leisure classes for pleasure and comfort during the long summer months. Among the latter was Beersheba Inn, built in 1839, and enlarged in 1875 by Colonel John Armfield.

Beersheba Inn was not all built at once. The methods of construction can be followed clearly, inasmuch as the first buildings were of logs of the one story "dog trot" type. The next group built shortly afterward was of handmade brick; the third and last development was the two-story "L"-shaped, columned building of clapboard siding which is now the major unit of the group.

The accompanying illustration is of the main two-story, pillared building and is typical of the inns of that period. The twin-room suites bordering the long veranda had their own pillared entrances so that every guest was, within the atmosphere of his own abode, perfectly at home. The balustrades are of light, interwoven wicker patterns of wood, which the Southerner developed delightfully on balconies and verandas. The tall, graceful, square pillars fade into one another as they nearly disappear into perspective, somehow never becoming tiresome even in their great number. They seem to say, "If one Colonel has a few pillars, many Colonels in a row should have endless pillars." And so is the functional honesty of the architecture a protection against monotony.

The ensemble is unique in that the plan of the building group takes the shape of a large rectangular court which furnished meeting places in the open for the guests. These gatherings might take the form of religious services, political speeches, or open air amusements. This rectangular shape also furnished a natural protection in the early days against the ravages of Indians and wild animals. It was in this court that Bishop Otey and Bishop Polk ministered to the Indians, and won their first Christian converts in that district.

In order to assist the proprietor to make Beersheba guests more comfortable, it was the habit of the stage coach drivers to sound their horns—one blast for each white traveler when they reached the bottom of the road leading up to the Inn. On their arrival at the top, about an hour and a half later, meals were ready for all the hungry guests.

One of the oldest and most substantial families of the Nashville community was the Peytons, whose names appear in the first expedition to the Cumberland country and in every public enterprise of antebellum days. In 1851, a Miss Kate Cunningham from New England, visitor in their household, accompanied the family to Beersheba Springs and wrote a vivid account of the trip:

"First and foremost rode Charles, the Colonel's intelligent and well-dressed serving-man; well mounted on a serviceable traveling horse, and leading by the bridle his master's noble battle-steed, which he still keeps as his favorite riding-horse. The horse is a large, finely formed animal, and with his gorgeous Spanish saddle half covered with silver, and his plated bridle, half of which was massive silver-chain, he moved

Grundy County, Tennessee

BEERSHEBA SPRINGS INN 1845

A typical mountain spa catering to distinguished Southerners of the Nineteenth Century.

on his way, tossing his head, and stepping off as if he "smelled the battle from afar off." Next came our family coach, a large, Philadelphia-built carriage, as roomy as one could wish, with drab linings, luxuriantly soft, broad, comfortable seats, that one could almost use as sofas. There were a dozen pockets in the sides, the two larger ones crammed for the occasion with books, magazines, and newspapers to read on the way when we should tire of each other, for the most social folks, with the most praiseworthy loquacity, can't always talk while traveling. One of the others was charged with cakes, and another thoughtfully teemed with peaches and apples, the foresight of the careful housekeeper, who had traveled with her mistress in her younger days, and knew how to make "white folks comfortable." A fifth, which was long and narrow, was neatly packed with cigars, to be conveniently in reach of the Colonel, the only smoker in our party; this care for making "white folks comfortable" being referrable to the attention of Charles, who was *au fait* in all things pertaining to his master's habits. A sixth pocket in the front contains a box of lucifer matches, to light the cigars with; and from a seventh projected the brass top of a small spy-glass, with which to view the distant prospects as we rode through the country. In each corner swung a brilliant feather fan, ready for our use, and in a rack over Isabel's head was a silver cup with which to drink from the springs or running brooks. There was an additional contrivance to the carriage I have never seen in any other; this was an arrangement by which the lower half of the front could be let down under the hammer-cloth and so make room for an extension of the feet of an invalid to recline at length; a luxury that the indolence of voluptuousness, rather than the comforts of indisposition, originated. Behind our carriage rode a little mulatto of fourteen, who is taken along as a pupil to initiate him into the mysteries of his future duties, as body-servant to the Colonel when Charles grows gray. . . .

In the rear of the carriage, at a sufficient distance to avoid our dust, and not to lend us theirs, rode on ambling nags, two female slaves, one of them Isabel's maid, who attends her everywhere, and Edith, who has been installed from the first as my factotum. It was useless for me to say that I did not wish to take her along, that I could do without her. Go she must, first because I should need her; secondly, she wanted to go and have the pleasure of the trip; and thirdly, Jane, Isabel's maid, would be lonesome without her companion to gossip with; and servants are better contented when they are together. So I had my maid. They were both dressed in well-fitting pongee riding-dresses, were mounted on side-saddles; and at the horns thereof hung the neatly tied bundles that contained their respective wardrobes. They paced along side by side after us, as merry as two young black crows in a corn field, and made the air ring with their mirthful and not unmusical laughter; for musical ever are the voices of the dark daughters of Africa; and I am not surprised to hear that there is a prima donna of this race in Paris, filling it with wonder at the richness of her notes. I can name half a score of negresses, on the estate of the Park, whose voices are charming, and, would, with cultivation, surprise and enchant the cultivated listener.

In the rear of these two "ladies," who only cease their talk with each other to switch up their nags, comes the coach-man's boy, a fat-faced, oily, saucy-lipped son of Ham, black and brilliant as a newly japanned boot. He is the coachman's page, and boy of all work about the stable and horses; and rubber down and harnesser-up; the polisher of the stable plate and the waterer of the horses; for your true "gentleman's coachman" is a gentleman in his way, and there are the meaner things of his profession which he leaves to the "low ambition" of such coarser clay as Dick. . . .

Dick was mounted on a low, black, shaggy Mexican pony, his feet dangling as if they were two weights to balance him and encased with a pair of brogans, the bottoms of which were still of that fresh polished leather-brown, which showed that they had not yet touched mother earth but were span new. . . . He led the bridle of Isabel's riding horse, the handsome creature I have before described, fully comparisoned and MY beautiful mule, accoutred with Mexican magnificence. These accompany us in order that, when we are tired of the carriages, we can ride, and also for our convenience while at the Spring. My mule is a perfect beauty! He is none of the Sancho Panza donkey race, but as symmetrical as a deer, with an ankle like a hind of the forest or like a fine lady's, with hide as glossy as that of a mouse, ears not too large, and well cut; a pretty head, a soft and affectionate eye, with a little mischief in it (observable only when Isabel would try to pass him) and as swift as an antelope, and thirteen and a half hands high. It comes at my voice, and does not like for any one but me to be in the saddle. You have no idea of the beauty and cost of these useful creatures in this country, and how universally they are used. Out of the nine private carriages at the Church last Sabbath, four of them were drawn by beautiful spans of mules. Even our own traveling carriage, which I have described to you, is drawn by a pair of large

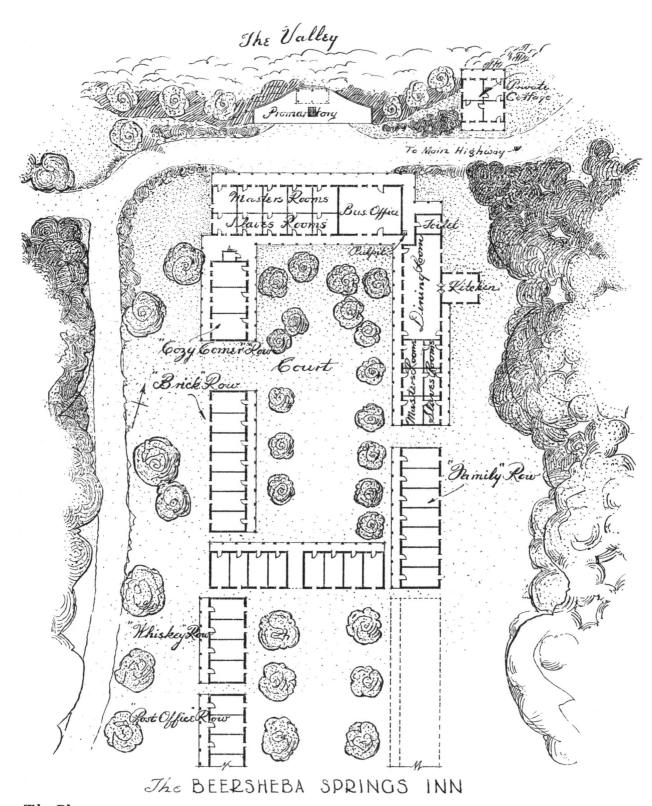

The Valley

Private Cottage

Promontory

To Main Highway

Masters Rooms

Maids Rooms

Bus. Office

Toilet

Pulpit

Dining Room

Kitchen

"Cozy Corner Row"

Court

"Brick" Row

Masters Room

Maids Room

"Family" Row

"Whiskey" Row

"Post Office Row"

The BEERSHEBA SPRINGS INN

The Plan

An open plan—a quadrangle of row houses, one room deep, with porches opening onto a common court, all covered with shade from enormous forest trees. Here one could relax in a combination of maximum privacy and comfort, scarcely noticing the difference between in and out of doors.

mules, sixteen hands high, and which the Colonel has been offered one thousand dollars for. It is only the rich that can afford the luxury of the use of these elegant animals. So don't smile at my saddled mule, which I have named "Jenny Lind."

I write to the measure of the dance in the hall, and the merry jingle of violins and castanets. The young folks are enjoying themselves while they are young. The happiest persons I saw in the ballroom, however, were the blacks. They stand in the doors and otherwise vacant places of the ballroom, and laugh and are as much at home as "Massa and Missis." They go and come around or across it as they please; a favored aunty will even ask you, "Please, Missis, stand dis way little bit, so I can see!" and "Missis" complies readily."

FAIRVIEW

NEAR Station Camp on the Nashville-Gallatin Turnpike, in the same neighborhood as the Peytons, are the lands of the Franklins, whose names in the middle of the nineteenth century were famous from Tennessee to the Orient as horsemen. Fairview, which I was about to visit, was a famous house of that clan. Tennessee folk today (apparently more than people elsewhere) find living in their ancestral homes to be a rare privilege. Seldom do we find one of these old houses vacant, or allowed to reach the stage of delapidation. So homelike, important and unmistakably private are their outward appearances, one hesitates to disturb the household by requesting an interview, the permission to photograph, sketch, or otherwise snoop into their peaceful situation.

I began to approach the house through a wandering lane bordered by walnut trees, the family mausoleum and other dependencies of more or less importance. At the end of the lane, Fairview loomed into sight, so silent on its slight elevation, I at once realized that the house was vacant. The feeling of interloper was not lessened over this new predicament; but accustomed to all kinds of situations in seeking out these old houses, I hastily considered my next move. Was the place barred? Or worse still, would I be unable to find someone whom I could question in the possibility of getting the local slant about its history and legends?

After a preliminary survey in which I had examined two elevations and found their openings barred, I heard sounds decidedly out of keeping with this scene of loneliness. Unquestionably musical (or rather unmusical) notes were coming from the direction of the east veranda. I investigated, and there perched on stone steps was a lone darkey endeavoring to extract music from a newly purchased guitar. So engrossed was this fellow in dreamy-eyed courting of the Muse that he seemed neither to notice my presence nor hear my questions. Finally I decided to share his musical efforts and, in order to gain an audience, offered to teach him a few chords if he would assist me in getting into the house. After about an hour of instruction and diagraming, with great concentration he was able to plunk the four major positions of the "G" chord. I then demanded that my part of the bargain be fulfilled; whereupon he motioned over his shoulder with a nod of his head and said, "I think de doe's unbarred—I don't belong here." Sure enough the door was unlocked and I took possession of Fairview for the day.

Standing back in order to take in the gigantic west façade, one is impressed by the graceful silhouette that combines two different but masterfully joined types of architecture. The original or main section is typical of that dignified form so prevalent in Middle Tennessee plantation houses. The featured entrance conventionally follows its contemporaries in that it is formed by a while panel extending from ground to roof in the center of the house, made up of first- and second-floor porches between two sets of superimposed, simple while columns and terminating in a delicate pediment. Its whiteness is exaggerated, especially in the late afternoon, when the setting sun illuminates the columns, pediments, and balustrade, bringing them out in sharp contrast with the soft salmon and red brick of the walls. To the right of the main section is the addition which consists of one long wing, stepping down in height as it extends southward. In contrast to the central portion, the wing is built in simple brick fashion suggesting the Spanish influence in the Feliciana country. There are broad arches containing carved wood balustrades behind which are deep and shady loggias serving the rooms. These loggias appear on both stories and continue across the front of the smoke house. Rich molded brick cornices are used at both first and second floor roof lines.

Inside is a large center hall extending through the house to another featured entrance to the east, duplicating the west one with even the treatment of the two-story west porch being repeated. The focal point of the center hall is the sweeping stairway separating the west and east entrances in such a manner as to form a front and rear entrance foyer. To the left of

Near Nashville, Tennessee   FAIRVIEW 1832

Isaac Franklin combined Nashville Basin stock and grain with West Feliciana cotton to create an economic empire of first magnitude.

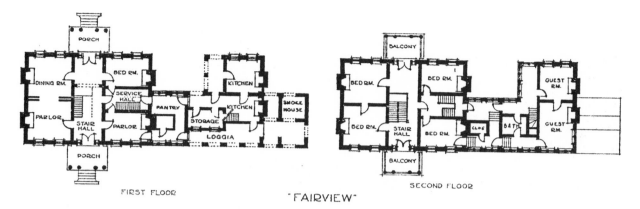

FIRST FLOOR   "FAIRVIEW"   SECOND FLOOR

the hall are the two parlors, characterized by the large, sliding doors separating them, high ceilings and marble mantels. To the right are the music room and dining room, separated by a secondary stair hall (added later) connecting the main section with the wing which contains the guests' rooms, vault, kitchens, and the smokehouse. Access to the rooms of the wing on both the second and first floors is by two long loggias on both sides of the house. The lower floor of the wing is given over to storage space and two large kitchens which facilitated entertaining on a large scale.

The second floor is devoted to guests' rooms and storage. The attic over the original section is plastered and apparently was used for bedrooms when necessary. It is interesting to note that on the walls of the attic there remain the names of many Federal soldiers, traced there in candle smoke at the time of their occupancy of the house during the War.

After passing through the house to view the remains of the gardens, outbuildings, race track, and barns, one immediately finds himself reconstructing a scene of teeming activity, ranging from slaves cultivating the expansive gardens through the vital work of raising crops, to the lavish industry of breeding and racing fine horses. Inevitably one visualizes the master of Fairview as a strong, ambitious man.

Isaac Franklin, the builder, son of a pioneer settler neither rich nor poor, was born on Station Camp Creek, in 1789. He was endowed with the spirit of conquest, and at an early age began to build a fortune. He was one of the first among Middle Tennessee planters to realize the lucrative possibilities in the raising and marketing of cotton. In this he anticipated the newcomers who were to flock into the South; men drawn there by their sudden realization that the climate and soil, together with slave labor, afforded great opportunity for those who could acquire the good land first.

Franklin's first purchase of land in the vicinity of Gallatin in 1831 comprised some two thousand acres, where he was able to build, a year later, the original section of Fairview. It was at that time conceded to

be the finest house in Tennessee, and as the years passed, he installed the many additional features which comprise the present house.

With the success which he enjoyed at Fairview came a desire to extend his operations afar. He went to Louisiana, where (I quote from *The Historic Blue Grass Line*) "in May 1835, he purchased a partner's half of nearly eight thousand acres in West Feliciana, upward of two hundred slaves, and all of the stock necessary for the immense plantation. He immediately formed a partnership with a resident of the Parish for the purpose of carrying on, as it was expressed, the business of planting upon several plantations situated in the Parish." A few years later, he became the "undivided proprietor of the vast plantations in which he was interested and had accumulated together more than five-sixths of his colossal fortune in immovable property."

In 1839 great changes took place at Fairview. A macadamized surface was put on the drive up to the house. More important by far, Isaac married Miss Adelicia Hayes of Nashville, daughter of a prominent citizen. While we do not know definitely, we presume that it was in approximately this year that the addition to the right of the house was made. Franklin needed more space in which to live and entertain in the style he had attained and which his wife had inherited, and it is reasonable to assume that the south wing was added after his years in the deep South.

By 1846, Franklin had acquired still more land in Louisiana until his holdings numbered seven large plantations in West Feliciana Parish and hundreds of slaves, beside his investment at Fairview. His death in that year left his widow the wealthiest woman in America. During his busy life he had found time for cultural pursuits, and his philanthropies were greater even than those of Cornelius Vanderbilt, benefactor of education in Nashville.

I never think of Fairview without wondering if the guitar-playing negro ever mastered the "G" chord, as I was never quite sure I knew it myself.

Nashville, Tennessee   BELMONT 1850

A great lady remembered days at Le Petit Trianon at Versailles.

BELMONT

THE traditional environment of culture that followed through this line of famous families and houses, beginning with the stirring rise to success of Isaac Franklin, reflects itself again in Belmont. His widow, Adelicia Hayes Franklin Acklin, carried on the tradition that had been her former husband's and, on moving to Nashville where her activities could extend more broadly, built another and even more elaborate home. Mrs. Acklin, descendant through her father's family, of the English Bishop of Bath and Wells was conspicuous among the Americans who added to the brilliancy of the Court of The Tuilleries in the closing days of the Empire. During her residence abroad, she absorbed much of the beauty and tradition of England, France and Italy and brought it back with her to Nashville. The influence of these countries can be seen in the architecture and gardens of her home.

Belmont was not the plantation type, as was the house Mrs. Acklin had recently left, but was designed to become the center of social life in a flourishing city. She furnished the house in rosewood and mahogany, made to order from the design of contemporary artists. Many of the fine pieces of statuary are still in place, such as the famous figure of Ruth, Randolph Roger's masterpecè, in the grand salon. Electric lights now flash from the heavy chandeliers of bronze and crystal almost worth their weight in gold; winter fires blaze beneath carved white marble mantels with tall mirrors above. Otherwise little is left of the original equipment.

The building is now part of Ward-Belmont, a school for girls, and has been altered for practical reasons. In the following discussion, therefore, we shall speak of it as it was originally built. The façade presents a formal composition, not unlike that of Belle Meade, the Hermitage, Merca Hall, and others, in that it consists of a large two-story central section with lower, nominally one-story wings on either side. These one-story wings are in reality two-story because the first floor of the central section is raised so high as to cause the basement to be completely above the ground. Insofar as the mass is concerned, the only departure of Belmont from this type is the addition of an observatory tower, which nevertheless completes the silhouette in a pleasing manner.

However, as soon as we note the architectural detail and the materials of construction of Belmont, all of its resemblance to Belle Meade, the Hermitage, and Merca Hall is lost. We see that here is a strong resemblance in fenestration, columniation, and detail to the Petit Trianon in the gardens of the palace at Versailles, and we realize that this little palace style has appealed again to the discrimination of a great lady. The main entrance consists of a recessed portico with two Corinthian columns forming three bays between the end projections that form the recess. These columns support a well-proportioned cornice, extending over the portico and continuing around the building. Superimposed above the cornice is a low parapet wall the corners of which are accentuated by marble statues. On either side of the entrance there is a delicate one-story projecting porch done also in the Corinthian order and surmounted by a balustrade. The roofs of these porches form balconies for large windows opening on to them from the second floor. A light, cast-iron balcony is contained within the entrance recess at the second floor. This recess and balcony depart in plan from the Petit Trianon, although the parti is not destroyed in elevation. A second and more important departure is the addition of the observatory on the top, which unfortunately is lacking in taste and refinement of detail, being surmounted by a crudely bracketed cornice.

The wings projecting on either side carry out in detail the same features as those of the central section of the house. Cast-iron balconies of a lace-like character have been added. These balconies have been copied, but do not occur elsewhere in this section so early, and much attention has been directed to the house because of them.

On entering the house, the visitor is admitted to a small formal foyer with a white, Carrara marble fireplace directly opposite the entrance and framed by two large doorways. Surmounting the two doorways on either side are transoms of red Venetian glass. This whole form of composition, in addition to being a physical obstruction, is also a psychological barrier against entrance to the great hall beyond. One is immediately reminded of the same treatment employed in one of the most famous houses of Virginia, Clairmont, on the James River.

Proceeding farther into the house, one passes under the grand staircase. This stair rises from the great hall, divides and passes over the two doorways leading from the entrance foyer. Emerging from under the stair, one comes out into the great hall. A fluted Corinthian colonnade borders the room on the stair side, supporting an elaborate plaster cornice of rich and almost baroque detail, which continues around the entire room. The hall is spanned by a vaulted ceiling, rising from the cornice and decorated only by a simple plaster ornament of leaf design. The chandeliers hang from the center of this ornament. Directly opposite the stair a large bay window projects to the rear of the house. On either side of this bay is a group of three arched windows set in deep panelled reveals. Of singular interest in this room is

Nashville, Tennessee   BELLE MEADE 1853

John Harding's contribution to the white-pillared house was inspired by the Choregic Monument of Throfyllus of the Athenian Acropolis. William Strickland, probable architect.

the absence of any apparent source of heat, for there are no fireplaces and no flues.

The two rooms on the left and on the right of the entrance foyer have the same combination of fireplace with doors on either side. They, too, are embellished with rich cornices, heavily molded doors, Carrara marble mantels, and great, gold-framed mirrors. The remaining space on the first floor is divided into rooms whose uses are not known, but which must have served personal or private functions. They are not treated with the lavish display of the great hall, but with restrained dignity.

The grand staircase to the second floor is characterized by slender white balusters supporting a mahogany rail, mahogany newels, dark oak treads and white risers and stringers.

Bedrooms comprise the major portion of the second floor, but its most interesting feature is the stair leading to the observatory tower. Reversing the system used at the lower floor, this little stair is divided at the bottom, rises in the form of an arch, meets at the midpoint and extends upward as a single stair. This stair hall is also treated with Corinthian columns supporting the arch over the stair well.

Belmont Gardens were designed in the manner of the Italian Renaissance of Tuscany, by an Irish landscape man named Mike Mullins—a cosmopolitan touch.

As one goes out of doors, he finds that careful workmanship and studied plan do not belong to the house alone. The garden which affords the setting for the lovely mansion was in its prime a spectacle of grandeur unsurpassed and almost unrivalled in Middle Tennessee. There is a succession of three large planting circles diminishing in size as they recede from the house in a series of lower levels. The garden stretches in circled formal beauty from the house at the top of the hill to the long nursery or conservatory at the lowest level. The first circle, near the mansion, is a formal lawn embellished with marble statues. Two tea houses are between this circle and the house, one on each side of the walls to the steps. A central walk divides this circle into two parts, and a simple marble fountain is in the center. Two smaller walks lead off from this circle terminating at two more smaller tea houses.

The center circle is divided by flower-bordered walks into three planting areas with points of intersection accented by boxwood. In the center of this circle is located the largest of the tea houses, which affords the only major break in the view from the house to the conservatory. The third and last circle is an involved mosaic composed of three circles with cross-walks, where colorful, less formal flowers bloom. In the design of this garden, the main point was to keep the flowers as near the ground as possible and to avoid too much foliage. Here, low plants give flat masses of color, producing the effect of a huge rug spread on a great green floor.

The other buildings in the Belmont group include the art gallery, bowling alley, ice house, zoo, gardener's house, green house, small tower, and stables. A lake, an orchard, and a deer park complete this estate, which was one of the most elaborate developments in the South.

BELLE MEADE

In 1853 John Harding built Belle Meade, the third house erected on his estate just west of Nashville. Since 1806, continued enlargement and development had been taking place until there were five thousand acres under stone fence. In addition to being first in many accomplishments, Belle Meade shared with Fairview the early local reputation of producing fine thoroughbred horses. It became the outstanding stock farm in America with an international reputation. Iroquois, the only American horse ever to win the English Derby, bore the colors of the farm, as did others whose winnings added prestige and riches to the Tennessee Bluegrass.

BELLE MEADE

Cradle of thoroughbred horses in America, Established 1839 by General W. G. Harding. Home of Virgil, Bonnie Scotland, Iroquois, Duke Blackburn, Enquirer, Proctor Knott, Great Tom, Tremont, Bramble and the Commoner. Site of Fort Dunham built in 1788 as defense against the Indians.

Erected by W. O. Permer, N.A.C.

As one stops to read this marker, he notices the stillness deepened by running water under the iron bridge. From this setting one glimpses the grey mansion, scarcely visible through the long avenue of cedars, on the top of a sloping hill. This initial view presents a restful study in shades of green and grey, with sprinklings of irises, marcissi, snow drops, columbine, and spirea drooping to the edge of the water. English ivy covers many tree trunks, offering a contrast to the numerous flowering trees of pink and white dogwood, lavender, white lilac and cherry.

About half way up, the drive is divided by a heart-shaped pool, outlined with a formal bed of small leaf periwinkle and star of Bethlehem. A huge circle starting at the pool is connected on the opposite side with a paved walk, flanked at the end with ivy-covered stone posts. Here, spreading from these posts is enormous boxwood. A wide bed of ivy ties the front of the house to the ground and, like sentinels on guard, stand two beautiful old holly trees. The first Harding is said to have brought with him a famous Swiss gardener from abroad to develop the park, and the formal beauty of the place substantiates this story. Its gardens and game parks contained several hundred acres along Richmond Creek.

The façade is graced by six enormous white square pillars of stone, which support an entablature of simple design. The pillars are each carved out of just two stones and are of excellent design. I strongly suspect William Strickland was the architect. Strickland, who was building the State Capitol at the time, was in possession of a copy of Stewart and Revett's *Antiquities of Athens,* whence came the parti of the state house. Among its precious pages is a measured drawing of the Choragic monument of Throfyllus, which was the inspiration for John Harding's house. Here we have Nashville's contribution to the Greek Classical tradition which had become the favored influence of the day.

The parapet with its break on each second pillar and crown in the middle seems to have some relation to the style of the Hermitage and Belmont. However, we find five Greek acroterea which adorn the parapet and which prepare us for other Greek detail. All doors and windows are heavily cased with pilaster and hood with Greek mouldings and ears, but there is otherwise little sign of the conventional Greek motifs usually adopted by the Revivalist builder of the middle nineteenth century. The center door is especially well proportioned. It is duplicated on the second floor where it opens on a small balcony, which has a cast iron railing. The roof of the portico is flat, but the main wing, which recedes from it is typically gabled in the Middle Tennessee manner with a parapet between large chimneys. A service wing flanks the right, most of which has been added in late years.

To General Harding also goes credit for improvements in the design of slave quarters. So much for the creator of Belle Meade—now for its benefactors.

To preserve the spirit of the South, it is most desirable that her old masions fall into appreciative hands, otherwise some of the most admirable examples of home and garden art will be lost. Belle Meade is fortunate in the hands of its new owners, Mr. and Mrs. Meredith Caldwell, whose ancestors have long been connected with Nashville.

Their contribution to General Harding's house is noticeable principally on the interiors. Upon entering the house one notices first the grace and continuity of the spacious stairway that winds three stories from the great hall. A warm color note is supplied from the Bohemian glass transom over the entrance door and the upstairs door leading to the iron-balustered balcony. The color scheme on the main floor, as well as of the winding stairway and second-story hall, is keyed to this glass, and the floors are carpeted with the same rich crimson in the manner of the eighteenth and nineteenth centuries. The wallpaper is a Creole pattern: white background with medallions of urns and roses in gray and ivory. Originally found in an old Louisiana plantation home, this pattern was blocked for Belle Meade. The woodwork in the entire house is white. Between the doors leading to the parlor is a Chippendale sofa upholstered in blue, flanked by a pair of rosewood commodes. Over the sofa hangs an oval French gold-leaf mirror, while over each commode is a family portrait. On the other side of the hall (as in the Hermitage) is another sofa, a Duncan Phyfe, upholstered in striped crimson and gold. Over this sofa hangs (so the owner told me) her most prized possession—a portrait of her father and her Aunt Ellen as children. Under the curve of the stairs is a Duncan Phyfe table on which is a lamp given to the family by Aunt Kate, a former negro slave.

The parlor runs the full depth of the house, with sliding doors at each end, typical of the trend of the day, to tie the indoors with the out. The side of the parlor which is broken by the two doors opening into the center hall has an enormous French mirror decorating the space between. On the opposite wall are two Adam mantels. Here the wallpaper is French Colonial, Louis XV, vivid medallions of deep pink and yellow roses, mauve tulips and a small trailing vine of blue flowers. The cornices are gold leaf. A pleasing glow throughout the downstairs is diffused through crimson window curtains, matching the carpet.

One piano (the only modern note in this otherwise Victorian room) is draped with the gorgeous colors of an ancient vestment, brought from Mexico a half-century ago. Another piano is more at home in its Victorian surroundings between the two mantels; this is of carved rosewood, made by Samuel Gilbert of Boston. The drawing room or parlor furniture, made by Belter, is carved rosewood upholstered in crimson. A handsome three-tiered table has a secret compartment. Another table holds a graceful Bohemian vase, and there is a love seat and a pair of Sleepy Hollow chairs of the same period as the Belter pieces. All in all, this room is Victorian par-excellence with none of the usually associated stiffness or stuffiness, and with all of the livable brightness and most charming aspects of the period.

In the manner of Victorian houses, this one has both a front parlor and a sitting room downstairs. This sitting room has a color scheme of blue, keyed to the large Persian rug. A comfortable sofa and deep arm chairs are inviting, and the mantel interest centers on a delicate gold French clock under glass (date 1872) between two dark blue and gold perfumers.

The dining room has wallpaper made for Belle Meade by Zuber of Paris. The dining table is oblong with carved apron and legs, and the twelve carved chairs have open-work backs and seats upholstered in striped red damask. A long sideboard has a white marble top and carved panelled doors and drawers with highlights in gold leaf. On the opposite wall is the white mantel with fluted pilasters, on which is a matched pair of hurricane shades protecting old candle sticks. Between the two windows facing the garden is a rosewood wine chest, its contours showing the Egyptian influence found in some Empire pieces.

Next to the dining room is Mr. Caldwell's study, containing one of the original massive Belle Meade mantels. A livable, masculine room with red leather chairs, deep wall bookcases and rare first editions. Here Theodore Roosevelt founded the "Boone Crockett Club."

The exigencies of diplomatic entertaining, for which the builder of Belle Meade was obliged to prepare, required ample space, hence the substantial old kitchen was of generous proportions. This kitchen is connected to the main house by a large enclosed porch. The pantry and laundry are paved with immense stones, six by four feet and two feet thick.

An old Austrian pattern of wallpaper in pale serene colors gives an air of tranquillity to twin bedrooms separated by great sliding doors. These rooms are so similar in furnishings, the impression is that the door opening between the two must be an enormous mirror reflecting the first room.

In two of the upstairs bedrooms are beds which were original Belle Meade property. One bed is most unusual in that it has a carved prayer rail on either side, padded and upholstered in light blue plush. On it is a priceless old crocheted bedspread. The other bedrooms are done in equally good taste.

And so a toast for the benefactors of Belle Meade: May they enjoy its gracious shelter for many more generations. The rest of us may rejoice too, for something fine and rich in American life is preserved here.

MARYMONT

MARYMONT, situated about three or four miles out from Murfreesboro in the heart of the horse country, is another charming old place characteristic of the Southern white-pillared mansion, and perhaps the last built in Middle Tennessee that conforms to the conventional ante-bellum type.

It is beautifully situated, far back from the road at the end of an avenue of pines, oaks, mimosa, and magnolia trees. Approaching the house from the road along the direct avenue, we catch only fleeting glimpses of the formal Ionic portico through the trees. Marymont was not the first house built on this property, as it replaced a crude pioneer structure, parts of which still stand. The land came into the possession of Aaron Jenkins, great-great-grandfather of the present owner, near the beginning of the nineteenth century as a reward for service in the Revolutionary war. The present structure was built about 1861, and it is said that the workers laid down their tools to take up arms in the Confederate Army.

The house is, in the main, typical of the period as to elevation and plan. A broad, flat façade is symmetrically treated with a large Ionic portico in the center and two windows at each side on both floors. The columns of this portico extend the full height to the cornice and support a pediment which does not conform to the dignified simplicity of the pure Ionic. The cornice in addition to the variation in mouldings is not improved by the use of heavy wood brackets.

This bracketed cornice at Marymont is among the few to appear in Middle Tennessee. The bracket does not add to the appearance, or conform to the accepted order of architecture, and there has always been a mystery as to its origin—architects invariably disagreeing. Some say it is the outgrowth of the modillion of the Corinthian Order combined with the triglyphs of the Doric. This combination is apparent on Vignola's so-called Cantilever Cornice at Caprarola. Another ancient form which may serve as precedent for the bracket is the distorted form taken by the triglyph of the fourth order of the Colosseum. Leicester Holland added this marginal note to an early proof of this manuscript:

"I think this bracketed cornice has nothing to do with Roman or Italian architecture, but is a 'Queen Anne' detail introduced from England and very popular in the Middle Atlantic States from 1860-1880."

Thus we find that, although the wood-bracketed cornice of the Southern home is severely criticized by purists and while it does not appeal to conventionalized thought, it did probably evolve from historical precedent, taking its particular shape because of the materials used.

Marymont employs the usual center stair hall with double parlors on the left and a reception room and dining room on the right. An interesting feature of the reception room is the tiny, partially enclosed private stair connecting this room with the one immedi-

Rutherford County, Tennessee

MARYMONT 1861

Those who built Nimrod Jenkin's house discarded tools in favor of guns—"The War-between-the-States had started."

ately above. Another memorable detail is the wall-paper in the front parlor, which is original and is said never to have been cleaned. It is a dead white, soft-textured paper, imported from France, and decorated with a simple pattern executed in gold leaf. Its excellent state of preservation and seeming freshness is remarkable.

Immediately to the rear of the major section of the house, the service wing extends back from the right side to form a long L. A dog-trot opening occurs in the service wing, separating certain of the service rooms from the others. The kitchen and cook's quarters are in this wing, at the end of which is the smokehouse.

Marymont is one of the few old houses remaining in the vicinity of Murfreesboro and, in the opinion of all who visit it, is the most charming example of old Southern architecture remaining in that immediate section.

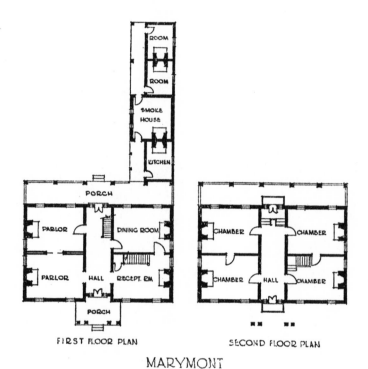

MARYMONT

Harrodsburg, Kentucky   DIAMOND POINT 1840

This is Kentucky. White-pillared houses here date from 1827 when architect Gideon Shryock, pupil of William Strickland, designed the State House at Frankfort after the manner of the classic architecture of ancient Greece.

PART TWO

KENTUCKY BLUEGRASS REGION

THE southern shores of the Ohio from the Big Sandy to the Mississippi have always been desirable land, easily accessible and richly varied. Thick forests clothe the mountains on the east, but westward there are open woods and grazing lands and barrens divided by great knobs and narrow streams. The region was so fair in the eyes of the Indians that they fought each other for the right to hunt there, and no one tribe dared to live upon it. Cherokees, Shawnees, and Chickasaws met Algonquins and other northern warriors in battle over the rolling land, rich in clover and bluegrass and over-run with game. Probably the name Kentucky originally meant "rich meadow lands," although tradition likewise gives it the meaning of "dark, bloody ground." Into this coveted territory of the Indians the first spearhead of American advance across the mountains could be deeply thrust because it was comparatively free of permanent Indian settlements.

Nature plays favorites in Kentucky. Through the middle of the state, extending in a northerly and southerly direction from the Ohio River to the Tennessee state line, rises a broad, low ridge exposing the limestone rocks that by their decay form the incredibly fertile area of Central Kentucky. To either side of this section lie other formations, some rich in minerals, but none so productive and so suitable for man's occupation. The soil is so fertilized by the limestone beneath it that there are fields which have been under steady cultivation for a century without exhaustion. Together with the Nashville Basin, a similar geological formation, it is the best soil in the Southland. The Bluegrass country takes its name from the color of the waving fields of grass seed at fruiting time—an unforgettable blue.

This was the land into which Daniel Boone led his companions in the spring of 1769. "Early in June," writes Theodore Roosevelt in *The Winning of the West*, "the adventurers broke through the interminable wastes of dim woodland, and stood on the threshold of the beautiful blue-grass region of Kentucky; a land of running waters, of groves and glades, of prairies, cane-brakes, and stretches of lofty forests. It was teeming with game." No wonder Boone, in spite of the perils and hardships of his trip, decided to possess this new land. "I returned home to my family," he wrote in his autobiography, "resolving to bring them as soon as possible to live in Kentucky, which I esteemed a second paradise, at the risk of my life and fortune."

News of the Kentucky lands spread like wildfire in Virginia and the other states east of the mountains. First came the hunters and explorers, then the surveyors, and then the axe-bearing men with their wives and children. Even during the years when Washington's soldiers were stubbornly fighting the English troops along the seaboard, many of the colonists heard the irresistible call to the west. It set them on the roads to Cumberland Gap where the Wilderness Trail led down into the new land, or to the forks of the Ohio to float down *la Belle Riviere*, as the French had named the stream. Although these settlers fought no battles in the Revolution, their bold advance toward the Mississippi was a part of the colonists' declaration of independence. Many of the place names tell how ardently the Kentuckians supported the cause of independence. Versailles and Paris, Fayette County and Bourbon County are compliments to their French allies. Lexington was named in honor of the opening battle of the Revolution.

In any migration the good lands were settled first. Naturally the rich Bluegrass region was early occupied, and by the beginning of the new century was a network of roads between thriving towns. Once the Indians were driven out, no hindrance remained to check its growth. Abundant crops of hemp and grain and tobacco, good horses and cattle, and well-balanced manufactories made for prosperity. Communication with the outside world was open along the Ohio and via the well-traveled traces to the east and south. Kentucky had a population of nearly seventy thousand when it became a state in 1792. By this time there were many "signs of stability and wealth" in the Bluegrass; and as always on an American frontier, the measure of that stability and wealth was the building of homes and public edifices. Log cabins gave way to wooden structures, and those in turn to brick ones. Stores, courthouses, and schools were erected.

Many Western travellers of the early nineteenth century recorded in their journals the culture and progress of the Central Kentucky settlements. In 1802, F. A. Michaux traveled along the Ohio Valley under the auspices of the Minister of the Interior of France. The route from Maysville through the Bluegrass is described in his *Travels*, and shows a keen eye for the customs, occupations, and homes of the people. Within a few miles of the start of his journey he noted "several very fine plantations in the environs, the land of which is as well cultivated and the enclosures as well constructed, as at Virginia or Pennsylvania," and a little farther along "two spacious inns, well built." In the town of Paris he observed that half the one hundred and fifty houses were of brick and that "everything seems to announce the comfort of the inhabitants."

In Lexington, too, Michaux found most of the houses of brick, and he approved of the many luxuries imported from England and France, such as muslins, tea, silk stockings, taffeta, and fine jewellery. As a scientist he was deeply interested in his conversations with Dr. Samuel Brown, a practicing physician in the town, who received regularly the scientific journals from London, collected fossils, and spent his scant leisure analyzing the mineral waters from Mud-Lick sixty miles away.

On his departure southward, Michaux spent five or six pleasant days at the plantation of General Adair on the Harrodsburg Road, recording the fact that the "spacious and commodious house, number of black servants, equipages, everything announced the opulence of the General."

An English traveller who was even more observant of architecture than Michaux was one Fortescue Cuming, who passed some weeks in the Bluegrass section in the summer of 1807. He, too, travelled down the Ohio to Maysville, then the "principal shipping port on the Ohio below Pittsburgh." He, too, fell in love with the beautiful countryside. "Perhaps there is not on the earth a naturally richer country than the area of sixteen hundred square miles of which Lexington is the centre," he wrote in his *Tour to the West.*

As he rode toward Lexington from Maysville he paused "at Col. Garret's fine stone house and extensive farm, where a young lady from an upper window, gave us directions." He called at the nearby home of the Baylors, hoping to hear the lady of the house perform on the piano-forte.

"Our ride now was on a charming road finely shaded by woods, with now and then a good farm, five miles to Johnston's tavern, where we fed our horses and got some refreshment. . . .

"The country continued fine, and more cultivated for the next six miles, hill and dale alternately, but the hills only gentle slopes: we then ascending a chain of rather higher hills than we had lately crossed, called Ash Ridge, we passed a small meeting house on the right, and Mr. Robert Carter Harrison's large house, fine farm and improvements on the left, separated by the north branch of Elkhorn river from Jamison's mill. We then crossed that river, and soon after, on a fine elevated situation, we passed general Russel's house on the right, with a small lawn in front of it, and two small turrets at the corners of the lawn next the road. The tout ensemble wanting only the vineyards to resemble many of the country habitations of Languedoc and Provence."

The two plantations referred to by Cuming are still among the landmarks of the region. The Harrison home, then known as Elk Hill, burned a few years after Cuming saw it, and was replaced by a frame house, still standing, which is known as Clifton.

This was once the home of the famous Carter Henry Harrison, five times mayor of Chicago. The Russell house was the two-story brick structure around which has grown beautiful Mount Brilliant, described later.

Cuming's fascinating account goes on to relate how he met with a gentleman on horseback near the Russell place who civilly entered into conversation with him regarding the condition of the region and of the nation at large. It was General Russell himself who obligingly rode some five miles of the way to Lexington with the stranger, thus inducting him into the rites of Kentucky hospitality. Shortly after parting with the general, Cuming had his first gratifying glimpse of Lexington, with its spacious, paved streets, handsome two- and three-storied brick houses, and well-stocked stores. A good supper and a pretty hostess at his inn completed his excellent reception.

There was no doubt that Lexingtonians were building in 1807. The new fifteen thousand dollar courthouse, then being finished was "a good, plain, brick building, of three stories, with a cupola, rising from the middle of the square roof, containing a bell and a town clock." There were also a "masonick" building, a public library, a university called Transylvania, and three creditable boarding schools for female education. Among other manufactories were jewellers and silversmiths, all prosperous; four cabinet-makers' shops "where household furniture is manufactured in as handsome a style as in any part of America, and where the high finish . . . is given to the native walnut and cherry timber"; and seven brick yards making annually 2,500,000 bricks. Cuming rightly counted it as a sign of progress that there were thirty-nine two-wheel carriages and twenty-one four-wheel ones in the community. And he noted a flourishing "place of fashionable resort" where a dancing master presided.

Like other early Americans, these Kentuckians expressed their personality and ambition in the homes and public edifices they erected. It was about the time of Cuming's visit that Henry Clay, a rising young lawyer lately married to Colonel Hart's charming daughter, purchased the place he named Ashland. It was, he declared, his Promised Land—as good a farm as there was in the world. With that combination of local pride and cosmopolitanism so characteristic of the Southerner, Clay engaged the famous L'Enfant, who had laid out Washington City, to plan his gardens, but he set out only native Kentucky plants and trees and shrubs. The plans for the house, of good Georgian influence, were drawn by Latrobe. Clay lavished money on Ashland to the end of his days—by buying marble and silver and books of hand-tooled leather that overflowed his octagonal library. Important persons who visited Lexington were royally entertained here by the town's most famous citizen. When Lafayette visited Lexington in 1825, he was fêted here, as were an impressive procession of presidents and senators and generals. As a matter of fact, so influential and rich had the Bluegrass region become, the master of Ashland need not have gone beyond his relatives and neighbors in order to throng the place with distinguished guests. They, in turn, would promptly have returned the courtesy under the roofs of their own well-built, spacious homes in the town or along the Richmond and Paris and Versailles Pikes. Then, as now, the Bluegrass people cared more for homes and estates than for public buildings and mercantile blocks.

Among the distinguished visitors at Ashland in the spring of 1818 was a scholarly young preacher from New England, Dr. Horace Holley, who had been "called" to the presidency of Transylvania University. In his memoirs, years afterward, Holley recorded the visit with Clay at Ashland. We can imagine that a conference with the already famous statesman in his book-lined library helped to persuade Dr. Holley that Transylvania could be made a great educational institution. Incidentally, a day with Colonel Meade at his beautiful estate, Chaumiere du Prairies, on the Harrodsburg Road, also made a deep impression on the young scholar. He doubtless concluded that such lavish hospitality augured well for the support of cultural projects. At all events, he accepted the presidency, which he held until 1827.

Transylvania was the first institution of higher learning founded west of the mountains. Chartered by the legislature of Virginia in 1783, firmly established in Lexington in 1789, its faculty included men like Dr. Samuel Brown, founder of the medical school, and the great scientist, Rafinesque. Perhaps its most brilliant period was during the presidency of Dr. Holley, who with his charming wife, entered fully into the life of the town. They were hosts to the intellectual and social leaders of Kentucky.

Morrison Hall is the most impressive building of Transylvania and one of the excellent examples of white-pillared edifices in the South. Any detailed consideration of this fine college hall lies outside the scope of the present book, but one can only with difficulty pass it by. The year after Holley's resignation, the trustees of Transylvania utilized the legacy of a Colonel James Morrison of Lexington to begin the erection of a hall to bear his name. Clay, deeply interested in every phase of cultural progress in the Bluegrass, advised the selection of the great architect Latrobe. The design, however, was finally drawn by a native of the town, Gideon Shyrock, the pupil of William Strickland, himself a pupil of Latrobe. Shyrock produced in Morrison Hall one of the noblest and purest examples of Greek Classical architecture in the South, worthy to be compared in beauty, if not in size, with Strickland's State Capitol at Nashville.

The Bluegrass region was one of the first transmontane settlements to become prosperous and stable; it continued without any serious backset to develop a deeply rooted culture. In its diversified agricultural system, tobacco, hemp, and stock were the money crops. At no time was it the beneficiary—or the victim—of such a sudden financial boom or tide of immigration as the cotton states of the Lower South experienced in the 'thirties and then again in the 'fifties. By steady infiltration of population and wealth, by its own rich products, and by intermarriage, the important families grew more and more permanent. Along the shady streets of the town rose residences like Glendower of the Wickliffes and Prestons, or Lindenhouse built by John Bradford the printer, or Hopemont, the home of the Hunts and Morgans.

Better still to the mind of a Kentuckian, the houses were often set in the center of rolling acres of rich land along the pikes that radiate from Lexington in all directions. There were Eothan (1798) built by the Rev. James Moore, and Winton (1823) home of the Merediths, and Woodburn House of the Alexanders, among scores of others.

Even the social upheaval of the War between the States failed to destroy the pattern of life in these Central Kentucky homes, although their halls rang with the spurs of hard-riding soldiers. On the threshold the sons of the house, some in blue and some in gray, kissed their mother and young wives goodbye. From the homes of the Breckenridges and the Clays went brothers who became leaders of both armies. There was no such divided loyalty, however, at Hopemont, from which gray-clad John Hunt Morgan rode out with five brothers to fight for the South. Emotionally and politically the Bluegrass was torn by the conflict; yet no march to the sea laid its homes in ashes. The ancestral houses of Central Kentucky were spared, and the community and family life in most cases survived the struggle. Today one is gratified to see many of the descendants of the original builders still living in these homes.

Not only geography plays favorites in Central Kentucky. Not only the unsurpassed fertility of its limestone basin, but its strategic location for transportation and trade, and even the fortunes of war have made possible here an orderly, unbroken development of culture rare in America. To these favorable factors must be added, too, the character of the early settlers and their descendants who have maintained their way of life as proudly as they have built their homes.

Lexington, Kentucky   WALNUT HALL 1842

Walnut Hall is the glamour type of Kentucky white-pillared house. Its stately facades with lace-like wrought iron accessories and statuary, traditional correctness of the orders, and rich detail, characterize the popular conception of early Kentucky Bluegrass aristocracy.

WALNUT HALL

NEAR Lexington, where the Bluegrass fields spread the farthest and the orderly plank fences fade on the horizon, you will find Walnut Hall, one of the oldest and most important estates, and the peer of any in beauty.

When I recall the first morning I drove out the pikes to see these homes in the Bluegrass region, I think of the miles of fences. Fences—those mute evidences of the character and temperament of the owner. Well designed and well kept, they proclaim orderliness in all things; neglected, they confess to carelessness in many things.

There is perhaps more freedom of design in a wooden fence than in any other; and the most attractive ones are found in the South, as well adapted to the present as to ante-bellum days. The Bluegrass region offers endless variations that have so impressed the public imagination that one cannot think of the breeding of live stock (particularly horses) without the Kentuckian's many conceptions of the wooden fence, practical, simple and sturdy, preventing stock from self-injury.

The beauties that lie beyond these gates at Walnut Hall have been greatly enhanced by the miles of driveway. There is probably no one feature on most already established estates so incorrect as the driveway, but here its design is all that can be desired from the aesthetic as well as the utilitarian point of view. As the stranger follows the drive through this great woodland pasture, there are surprises at every turn—dozens of fine barns and stables, many old homes for important employees, such as managers, bookkeepers, and technicians, and the rambling picturesque place of Harkness Edwards, son of Dr. and Mrs. Ogden Edwards, owners of Walnut Hall.

The grand scale of the mansion house is everything one would expect after the impressive "build up" of the approach through the estate. Walnut Hall is a fitting symbol of the glamour of the Kentucky Bluegrass country. It has historic background and tradition together with architectural correctness in scale, silhouette, detail and type. It has often been chosen to represent Kentucky homes in pictures and stories, and surely few Kentuckians could object.

The main façade of Walnut Hall presents a central portico crowned with two tiers of balustrades building up to the cupola, the whole forming a central motif of the silhouette. Then the elevations diminish as the less important parts of the house plan work toward the two one-story porches flanking the end pavilions. Each of these porches has a separate entrance to the garden. A departure of note is the color scheme; it is of dull yellow-painted brick with white pillars and green shutters. In plan there is the original square arrangement with center hall flanking rooms on each side. The later additions consist of balanced terrace wings, each with its own porch.

The picturesque quality of Kentucky houses is best epitomized in their porticos. Walnut Hall is a notable example of dignity and hospitality. It is interesting to compare it with Rose Hill. For this purpose both houses have been sketched from the same angle. The order employed at Walnut Hall is Doric; baseless shafts (not original) spring from a porch level four feet above the gardens and continue through the second floor line (which is evidenced only by a simple door and wrought iron balcony) to the entablature above. An increased classical effect is gained by a full proportional entablature consisting of architrave, frieze, and cornice in interesting decorative mouldings, and surmounted by a wood balustrade instead of the customary pediment or occasional hipped roof.

The Walnut Hall portico scores another triumph in its combination of metal designs in wrought iron and bronze with the heavy Greek classical motifs in masonry and wood. Because of the portico floor height it is necessary to have protection at the edges and steps. A happy solution in metal displays an unusual rhythm in circles—little circles, big circles, circles in base rails, top rails and field—that somehow escape becoming monotonous. Bronze statues form a fitting terminus for the step railing and are used as light standards.

The brick walls are made heavy and substantial by brick pilasters of equal proportion with the columns, which makes for better balance between walls and massive entablature. After leaving the portico, the entablature is carried without variation or alteration completely around the house, accompanied by the same balustrade.

On the interior, there is a winding stairway occupying the center of interest in the stair hall. To the left, double drawing rooms have walls of pale yellow satin damask with gold cornices. The furniture is upholstered in flowered patterns of satins. There are French cabinets of outstanding quality, pier mirrors and oriental rugs, along with a wealth of paintings adorning the walls. In the library one finds brocade-covered walls forming a background for numerous family portraits. There is a large living room and an interesting den with oak panelled walls. The large dining room has walls and ceilings of panelled mahogany. Here the furniture is late Georgian, and the floor covering a Royal Kirmanshah rug.

All in all, the interiors of Walnut Hall are pleasingly different from those of the average Kentucky house.

Fayette County, Kentucky

MOUNT BRILLIANT 1807

When Fortescue Cuming saw General Russel's house in 1807, it was a glorified pioneer type of simple Georgian influence. As the 19th Century wore on, the plan increased and the portico was added. It appears today as it was at the end of the white-pillared period—1861.

Mount Brilliant, like Andrew Jackson's Hermitage, grew through several periods from glorified pioneer house to white pillars and to this day it vividly reflects the social and economic life of those times. It is a choice example of the simple white-pillared big house type of Kentucky.

It has been one hundred and thirty-one years since Fortescue Cuming in 1808 pointed out "General Russel's house on the right." It has now been the home of the Louis Lee Haggins for more than twenty-five years. Because of their thorough understanding of the Southern way of life, they have become a part of Mount Brilliant, and today one cannot speak of Mount Brilliant without at the same time speaking of the Haggins .

The building of Mount Brilliant was begun soon after the colonies won their independence, when William Russell and Robert Spottswood Russell claimed the land secured by their uncle, Henry Russell, a native of Culpepper County, Virginia.

Like Walnut Hall, Mount Brilliant is a typical two-storied Kentucky white-pillared house, but be-

Lee Haggins knew it and this is why they chose to cast their lot with the Old Russell Place.

Originally the central hall had only one room on each side both upstairs and down, and there was no portico. Today, in addition to the portico, there is another room width on each end of the house which gives a gracious spread to the façade.

The main theme of the home is to tie simplicity with comfort, and one thinks in terms of log-sized fireplaces, broad floor boards and a complete accord between house and grounds. The stairway, as usual, is the first interior detail to meet the eye. The staircase probably reflects the influence of authentic architectural precedent more than any other of the interior details of a house. In this case one sees an example of a particular historical period, the eighteenth century. The first tendency to curve the stairway in the southern-most states was about the turn of the century—first in railing and landing and then in the run of steps.

On the walls of this house is a series of most interesting and valuable pictures by Ben Ali Haggin,

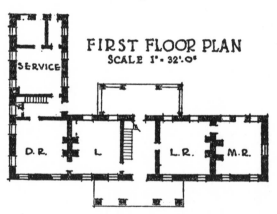

FIRST FLOOR PLAN
SCALE 1" = 32'-0"

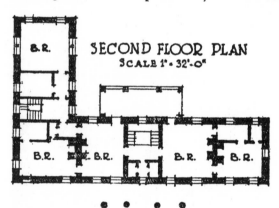

SECOND FLOOR PLAN
SCALE 1" = 32'-0"

MOUNT BRILLIANT

yond this, there is no similarity. One is glamourous—the other is simple, robust, indigenous. Gone are the stone proportioned entablatures with ornate Greek fretted mouldings, crowned by lacy balustrades; gone are the frail wrought iron balconies and stair rails, the rich pilastered brick walls, the statuary bronze torch standards and accessory porticos on each end. Mount Brilliant expresses a more modest simplicity in its portico of Greek Doric "Vignola" proportioned shafts and simple entablature and pediment. A linteled doorway is inconspicuous except for its two free-standing columns. There are plain walls and fenestration accented only by divided lights and green shutters. All is crowned by a hipped roof expressing a mass silhouette of simplicity, though sturdy and true. This is Kentucky as James Ben Ali and Louis

whose paintings contribute most pleasingly to its success. A grandfather's clock and a handsome old mirror over an inlaid half-circular table give the background one has come to expect in the stair halls of the South. To the right is a spacious living room which opens to a men's room, or "office." In the living room one is greeted with a formal outlay of fireplace and balanced "in between windows" treatment of period furniture and good art, the like of which few homes in America or abroad can boast. The office, as it is called, is in reality a sports' library. Here are amassed the family trophies and valuable collection of sports literature dating from early American history. On the left of the hall is the library and beyond, the dining room. The library seems to be the family room, for here, more than anywhere else,

Lexington, Kentucky   ROSE HILL FROM 1800

The rambling one-story houses with classical porticoes were favorites in Kentucky. Rose Hill House grew through formative periods of Kentucky life and was finally crowned with the portico.

we find pictures and trophies more closely associated with the prideful accomplishments as well as the tragedies of this interesting household.

In the dining room is a background keyed to the furniture, to the mode of living and entertaining, and above all to the personality of the hostess, Mrs. Louis Lee Haggin. Here is a fine example of dignity combined with friendly intimacy. The familiar Kentucky "drinking board" was not denied a place in this dining room. A Sully portrait of William Ransom Johnson, "The Napoleon of the Turf," is an over-mantel decoration, while Bohemian amber glass on the mantel and in a carved corner cupboard add generally to the color of the room.

On the second floor are four bedrooms, but the point of interest is without a doubt the children's library. This is a noteworthy contribution to the early development of the children of the Southern family. And surely, at one time, a governess held classes here between ten and twelve every morning.

From the back hall door, one notes that the connecting link between the house and the interesting grounds in the back is a brick terrace. This belongs equally to the house and to the garden—a successful treatment which should be always developed if possible. A walk leads to the formal evergreen garden, which is bordered by a high hemlock hedge, and is entered through a wooden gate, painted a pale greenish-blue, matching two other gates in the connecting garden. Old-fashioned flowers of all colors edge beds in front of the shrubs. Much may be said of the effective use of color throughout this garden, which is reflected in a large rectangular swimming pool, lined with the same soft, greenish-blue tone of the gates. The clumps of blue lily-of-the-Nile at the head of the pool surpass, to my mind, any other edging plant for this spot.

A glimpse through one of the gates of this walled garden carries the eye to the guest house—thought to be even older than the "big house."

The difference between the natural and the geometrical style in landscaping at Mount Brilliant is worthy of notice. The most finished place in the natural style, when neglected, soon ceases to be recognized as a work of art. But so long as a row of trees, or a terrace remains, it bears the stamp of art and proclaims itself to be the work of man. Mrs. Haggin says this is the reason for her box-bordered paths, hemlock hedges, brick and stone walls, concrete swimming pool and brick terraces. She wants the future generations of Mount Brilliant to carry on in the manner of the present. Hence a permanent pattern for this Bluegrass region of Kentucky.

ROSE HILL

KENTUCKY's contribution to the white-pillared house is, as is its cultural development, second to no other section of the South. I would choose, however, its picturesque one-story houses as an outstanding architectural accomplishment. There are hundreds, but some are especially commendable: Rose Hill, Mansfield and Botherum in Lexington; Springhill and Farmington in the environs of Louisville; and Grange in the vicinity of Millersburg, and Elmswood Hall in Ludlow. A worthy representative for purposes of illustration is Rose Hill.

Most Kentucky one-story houses have a complicated, inconsistent floor plan. Rose Hill is no exception. The room arrangement of a typical Kentucky two-story house developed along the same lines as in Tennessee (center hall and two rooms on each side, duplicated on the second floor). The one-story plan was as simple to begin with, but from the time the glorified provincial plan type added another story for bedroom, the one-story plan, if it were to remain so, had to devise various other schemes to take care of expansion.

The original central hall, flanked by rooms on each side, is recognized in most instances if the plan is carefully examined; otherwise it might not be noticed. In Rose Hill this arrangement is present but only in part, as the drawing room occupies the span back of the hall. Careful examination of the plan will convince one that here, as well as in most of these houses, the plan is not as practical from the standpoint of light and ventilation as the two-story plan. Kentucky, however, is farther north and these elements are not as imperative as, let us say, in the Deep South. At Ingleside, in Mississippi, the bedrooms are attached to the central hall and parlor and dining room by covered passages; at the Eslava Home in Springhill, Alabama, they are attached to the rear and have separate porches; while at the Beauregard house in New Orleans, they line up on the other side, with a long hall extending the depth of plan.

Getting back to Rose Hill, we find a court inserted on either side of the center living arrangement, which would indicate much concern was felt for this bedroom problem as well as for circulation and the general utility of plan. My theory is that the original plan of Rose Hill was a cottage with a central hall and four rooms. The room back of the hall, two end wings and portico were added later.

The street front is the façade we all admire, and features the type of portico Kentucky had a way of putting so much dignity into. Here it looks as important and dignified as the great two-story porticos of it neighbors. The wings, lower and unimportant, flanking each end give it additional "build-up," as the very fine Georgian doorway ensemble gives it a background and a cause for being. The entire façade, as in all one-story houses, depends much on silhouette to make for a successful plan portrayal and architectural appearance.

The order employed is Ionic and the proportions are carefully selected for wood construction. Other materials of construction are almost forgotten because

on the left of the hall is crowned with a pair of interesting French prints. This grouping balances nicely with the Victorian sofa on the opposite side. The framed coat-of-arms gives color to the soft tones on the Colonial scenic wall-paper, as do the valuable old Italian prints by Bertote. The highest praise for appropriate furnishings must go to the grandfather's clock.

In the living room on the right the fireplace

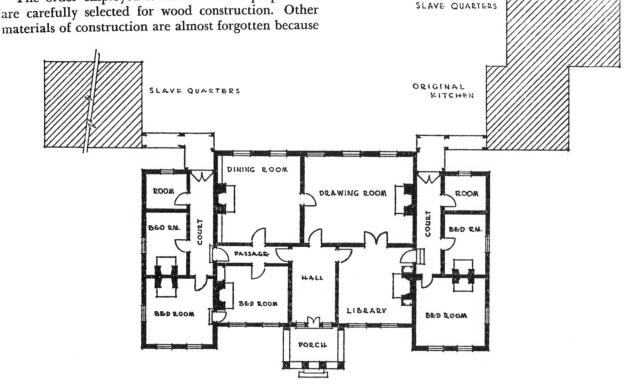

FLOOR PLAN
ROSE HILL

they seem to be just the right choice, one by one and in combination; we shall sum them all up by saying, "brick, wood and stone at just the right places."

From the quaint old lilac-bordered walk of octagonal brick to the wrought iron lanterns flanking the door, the place bespeaks hospitality. There is a picturesque air of informality which is due to the delightful personalities of the host and hostess, Dr. and Mrs. Davis G. Buckner.

Rose Hill was built by John Brand, about 1820. One of the first hemp manufacturers in America, John Brand had a very definite sense of duty to this community and countrymen, which will carry a lesson to all who have time to look him up in the historic documents of Kentucky. It is enough to record here that the present owner's forebears financed John Brand in his early American undertakings before the days of Rose Hill House.

And back to Rose Hill House. The flap-top table

ensemble consists of an excellent cabinet mantelpiece with overmantel decoration of a Jouett portrait of Tobias Gibson (an ancestor of Mrs. Buckner and founder of Jefferson University at Port Gibson, Mississippi). A pier mirror is on the front wall between two windows. It balances a very interesting fan-topped double door on the opposite side of the room, which leads into the drawing room. Reflected herein is much of the beauty of the room: early American vases, candle-sticks with their crystal prisms, ottomans, sofas and colorful Oriental rugs.

Opposite the front entrance and opening on the court in the back are the drawing room and dining room which contribute to an outstanding air of richness and good living. There is added dignity and refinement in the simple plaster panelling of the walls done in hydrangia blue. This lyrical color is further enhanced by the crimson upholstery and the long sweeping draperies that hang from gilt-metal cornices. Carefully planned arrangement of furniture and equipment makes the elegance of this room so warmly felt that it is impossible to pass without a word in its favor. Around this room, there is a generous sprinkling of warm red Venetian glass, which is banked in the old cabinet in the dining room. The hydrangia blue is picked up again in the blue bonnet in a portrait over the mantel. Flanking the mirror over this Egyptian marble mantel are two rose bottleneck vases. The desk at the left of the fireplace is an interesting example of marquetry, and the tip-tilt Sheraton table at the right has above it small paintings said to be Titians. A delicate little cabinet is elegant with its delicate legs and soft polished surface. The bedrooms, furnished in very old furniture, are as interesting as the rest of the house, but space does not permit me to dwell on these. I will say, however, that they are a triumph of restfulness.

The back court and rambling grounds in their different elevations create a distinctive atmosphere. These grounds, like the house have solidity; the one complementing the other.

Columbus, Mississippi

THE PRATT HOUSE 1850 (APPROX.)

A secondary entrance portico here displays an interesting arrangement of square pillars. There are other porticos **also**.

CHAPTER III

ALONG THE NATCHEZ TRACE

CHAPTER III

Along the Natchez Trace

SHORTLY before the American Revolution, when sturdy Virginians and Carolinians were casting hungry eyes on the rich lands over the mountains, a vast and vaguely defined territory to the Southwest came, as a result of European diplomacy, under the control of His Britannic Majesty. This region, known then as British West Florida, comprising about what is now southern Mississippi, southern Alabama, and the portion of Louisiana east of the Mississippi River, had early been explored by the Spanish and then controlled by the French, from 1699 to 1763. French capitals had been established along the coast at Biloxi, Mobile, and New Orleans; and an outpost, Fort Rosalie, was built on the bluffs above the river. In 1763 the British erected another fort on the site of old Fort Rosalie, and English-speaking immigrants began to pour in from the eastern and southern colonies. The result was the frontier town of Natchez.

Many factors combined to make Natchez important. The spot was originally occupied by a powerful tribe of Natchez Indians, who realized the value of its strategic location on the high ground along the shores of the great river. Under French rule it was a trading post of importance; and under British occupation a border town and an outpost of English influence against Louisiana. In 1783 the Spanish, coming over from East Florida, gained control of Natchez for a time, and opened it to American settlers on attractive terms. When the United States gained the district under the Treaty of 1795, Natchez became the last important American settlement down the river for the boatmen with cargoes of corn and pork for the foreign port of New Orleans. It was the southwest corner of American power. Moreover, the little town was prosperous in its own right as the center of a good farming district.

The early settlers in Natchez included well-to-do men, many from New England, who gave the place a permanently aristocratic stamp. Later comers continued to be predominantly British even during the Revolution, during which period many Loyalists came out to this remote territory where the war could be all but forgotten in the struggles of frontier life.

Yet Natchez was more cosmopolitan than other American frontiers had been, for besides the Indians who still inhabited the interior, there were French and Spanish residents, not to speak of the half-breeds. What lent the settlement its distinction was the air of culture which the original settlers brought from the older east and never surrendered.

Natchez was a point of interest to every traveler. A most discerning view of it is that of Fortescue Cuming who perceived even at that early date that the strains of American, Spanish, and French blood met and mingled with striking results.

"I was much struck with the similarity of Natchez to many of the smaller West India towns, particularly St. Johns Antigua, though not near so large as it. The houses all with balconies and piazzas — some merchants' stores — several little shops kept by free mulattoes, and French and Spanish Creoles—the great mixture of colour of people in the streets, and many other circumstances, with the aid of a little fancy to heighten the illusion, might have made on suppose, in the spirit of the Arabian Nights' Entertainments, that by some magic power, I had been suddenly transported to one of those scenes of my youthful wanderings. . . . On Thursday, the 25th, I arose early, and sauntered to the markethouse on a common in front of the town, where meat, fish and vegetables were sold by a motley mixture of Americans, French and Spanish Creoles, mulattoes and negroes. There seemed to be a sufficiency of necessaries for so small a town, and the price of butcher's meat and fish was reasonable, while vegetables, milk and butter were extravagantly dear. . . . Proceeding to the southward from Natchez, in two short miles I came to Colonel (late governor) Sergeant's handsome brick house. The road led through a double swinging gate into a spacious lawn, which the Colonel has formed in the rear of the house, the chief ornament of which was a fine flock of sheep. The appearance of this plantation bespoke more taste and convenience than

I had yet observed in the territory."

Indeed, Natchez had been on a main route of travel up and down the river long before the coming of the white man. From prehistoric days, the Indians went down the swift river in their canoes with ease only to find the trip back too difficult to undertake. The strong currents around the hundreds of bends in the Mississippi forced them to seek other ways to return. The numerous tribes from the Ohio Valley country usually left the River at the bluffs of Natchez striking off the northeast along the watershed of the Big Black River and the headwaters of the Yazoo and Tombigbee, crossing the Tennessee below Muscle Shoals, and pushing on past the Cumberland hills toward the hunting grounds along the Ohio. This trail known after the coming of the whites as the Natchez Trace, was a narrow, hard-beaten footpath which, crossing streams, swamps, and forests, led up from Natchez to the settlement on the Cumberland which came to be known as Nashville. Soon the white settlers learned that it was better to travel by land than to fight upstream against the Mississippi; and very early the boatmen began to dispose of their craft in "Orleans" or Natchez and to walk back along the Trace to their homes. So essential was this route that in 1801, the United States Government bought the right of way from the Chickasaw and Choctaw Tribes and began to construct a road.

The following year, the indefatigable F. A. Michaux described the Trace:

"The road that leads to the Natchez was only a path that serpentined through these boundless forests, but the federal government have just opened a road, which is on the point of being finished, and will be one of the finest in the United States, both on account of its breadth and the solidity of the bridges constructed over the small rivers that cut through it; to which advantages it will unite that of being shorter than the other by miles. Thus, we may henceforth, on crossing the western country, go in a carriage from Boston to New Orleans, a distance of more than two thousand miles."

The expenditures of the government on the Natchez Road seemed immense, amounting to six thousand dollars. The specifications called for the twelve foot road to be cleared of all trees, logs, and brush and made passable for wagons. All streams not over forty feet wide were to be bridged. The Indians were to continue the operation of the ferries along the shores for a long time. In spite of all its improvements, however, the road passed through unsettled lands, and offered the prospect of hardships even to stout travelers at the begining of the century. Alexander Wilson, eminent ornithologist, made the trip from Nashville to Natchez overland in 1810.

"I passed few houses today; but met several parties of boatmen returning from Natchez and New Orleans, who gave me such an account of the road, and the difficulties they had met with, as served to stiffen my resolution to be prepared for anything. These men were dirty as Hottentots; their dress, a shirt and trousers of canvass, black, greasy and sometimes in tatters. . . . These people came from the various tributary streams of the Ohio, hired at forty or fifty dollars a trip, to return back on their own expenses. Some had upwards of eight hundred miles to travel. . . . This day (Wednesday) I passed through the most horrid swamps I had ever seen. These were covered with a prodigious growth of canes and high woods, which, together shut out almost the whole light of day for miles. . . . The next day I passed through the Chickasaw *Bigtown*, which stands on the high open plain that extends through the country, three or four miles in breadth, by fifteen in length. Here and there you perceive little groups of miserable huts, formed of saplings, and plastered with mud and clay. On the fourteenth day of my journey, at noon, I arrived at this place (Natchez, Mississippi Territory), having overcome every obstacle, alone, and without being acquainted with the country, and, what surprised the boatmen more, without whiskey."

For some years the Indians were sporadically troublesome, while a great danger lay in the gangs of unscrupulous white men who hid out in the swamps and lonely wooded stretches to rob and kill travelers, especially those coming up from the south with money in their pockets. Even the tough flatboatmen sewed their hard-earned silver dollars in their buckskin breeches and went in parties for protection. These desperate land pirates were the remnants of the powerful bands of river pirates who had infested the Ohio a generation earlier until they were driven from their hiding place at Cave-In-Rock by the Kentucky militia. Then they organized under bold leaders like Tom Mason and the Harpes to rove over southern Tennessee and northern Mississippi, preying upon unwary travelers along the Trace. Finally, the Governor of Mississippi Territory offered a large reward in gold for Mason's head. Two men turned up with a bloody head, claiming the gold; but a mail rider who knew Mason denounced them as impostors; and the two men were themselves tried for murder and hanged.

In spite of all obstacles the mail from Nashville to the Southwest went regularly over the Natchez Trace. The post carriers, hardy enough to defend themselves, had less to fear than others on the road, for the robbers usually wanted only money. The rider left Nashville having tightened the leather mail bag to his saddle, adjusted the sack of corn for his horse, and his own blanket and provisions, and hung a bugle from the saddle horn to blow his signal of arrival to the settlements along the way. A wave to the crowd of onlookers, and he was off. Fifty miles to Gordon's Ferry and over Duck River. A fresh horse and on to the Tennessee, where the Indians would put across at Colbert's Ferry. On through the Chickasaw Nation and among the Choctaws . . . until at last he would sight the hills of Natchez and blow his bugle for the last time that trip.

Dangers lessened on the Trace after Jackson's men, victorious at New Orleans, marched homeward over it. Accommodations gradually became better. Where in 1815, a lady who made the trip up to Nashville in her carriage was compelled to camp out or to find lodgings with settlers along the way, her sons and daughters were able to stop at inns. If some of them proved a little uncomfortable, the sons at least could have solaced themselves with the memory of the incomparable mint juleps served at the Mansion House, prime hostelry of Natchez. The portrait of one such liquid masterpiece has survived in the writings of an early Mississippian.

"On Saturday we were presented with a magnificent Julep from the Mansion House that probably excelled anything of the kind made on the continent of Columbus. . . . It was in a massive cut goblet, with the green forest of mint which crowned it frosted over with sugar snow, and the whole mass underlaid with delicate slices of lemon piled in the pyramid of ice. As for the liquor, it was so skillfully compounded that no one could detect its several parts. Ladies drank of it and supposed that some huge grape from the south side of the Island of Madeira had burst open on a sunny day and been crushed in a goblet."

And he adds that at the Shakespeare, an hotel opposite the Market House, every julep had a moss rose in it, while the Steam-Boat Hotel put a strawberry in each one.

A colorful pageant was to be seen up and down the Trace in the first half of the last century. As the ranks of the boatmen and coonskin hunters thinned out, carriages became numerous, and were followed by black servants on horseback, evidences of the growing wealth of the region. Henry Clay traversed the thoroughfare as he campaigned three times for the presidency. Jefferson Davis, officer in the United States Army, and statesman, took the lovely Varina Howell as a bride over the Trace to his plantation home. Rich planters' sons went up the Trace to attend Transylvania College. Men with money to invest came south over it to seek out the fabulously rich cotton lands by which they could treble their investment in a couple of seasons. Wherever the road crossed the rich river lands, communities were established; Nashville and Maury County in the Cumberland Valley; near Muscle Shoals in northern Alabama; along the Tombigbee, the Black and the Yazoo. Communities like Columbia and Florence and Columbus and Port Gibson flourished in the decades before the War, and bore witness to the rich stream of culture that spread along the route of the Natchez Trace.

After the steamboats established upstream traffic, the actual importance of the Trace as a road diminished, but not the far-reaching contacts which it had fostered. In the fertile regions that it had tied together in commerce and culture, from the Bluegrass down to Natchez, a pattern of life had been established, based on plantation economy and expressing itself most characteristically in architecture. As in Middle Tennessee a little earlier, handsome homes took the place of pioneer dwellings. Where cotton was the dominant crop, wealth piled up with unbelievable rapidity, and a lucky planter stepped from a cabin to a mansion in four or five years.

The foundation of the ante-bellum plantation system, of course, was slavery, and its manifest advantages seemed to justify it. In 1840 Alexander B. Meek, an enthusiastic Georgian, wrote in *The Southern Ladies Book*:

"In a few years, owing to the operation of this institution upon our unparalleled natural advantages, we shall be the richest people beneath the bend of the rainbow, and then the arts and sciences, which always follow in the train of wealth, will flourish to an extent hitherto unknown on this side of the Atlantic."

However limited the results in the other arts and sciences that appeared in the cotton belt before 1861 to support Alexander Meek's contention, at least in regard to architecture he hardly exaggerated. First fruits of Southern prosperity were the ample, admirable homes that were built by the planters, sometimes on their estates, sometimes in the towns that were their headquarters. Many of them still stand and serve to make a trip today along the old Natchez Trace an exciting excursion into history and architecture. We shall start southward from Nashville, just as the mail riders did a century and more ago.

Franklin, Maury County, Tennessee

CARNTON 1830

Randall McGavock and his architect came from Williamson County, Virginia, to build his house. He called it Carnton after the family estate in County Antrim, Ireland.

PART ONE

TENNESSEE COUNTRY

County loyalty was a very real bond of unity in the early South. Even a large family would have most of its holdings within the boundaries of a county and would thus focus its social and political influence. In rich agricultural sections a vast local pride developed. For example, a planter residing in the town of Columbia would as a matter of course register at a Nashville hotel as being from Maury County, Tennessee.

Travelers along the Natchez Trace in 1830 or thereabouts were certainly aware that they were traversing a rich and delectable land. As it leaves Nashville, the road winds southwesterly through Williamson and Maury Counties, a section which shared the same pioneer heritage as the rest of the middle part of the state, but with a more direct contact with the Indian frontier. The first town along the route in the early days was Franklin, county seat, and already a thriving community.

CARNTON, THE RANDALL McGAVOCK HOME

A famous old home in Franklin is Carnton, built by Randall McGavock more than a century ago. When McGavock, scion of a Virginia family, migrated to Tennessee in 1824, he purchased a thousand acre tract in Williamson County and made plans to begin his mansion at once. He proposed to erect an ample establishment of more than twenty rooms, with smokehouse, workrooms, carriage house, and to call it Carnton for the family estate in County Antrim, Ireland. Family records show he had brought with him from Virginia for that purpose an architect by the name of Swope, to whom doubtless may be attributed much of the classical air of the house.

The wing pavilion plan, the façades of gables with twin chimneys tied together with the parapet wall above the roof lines, as well as the porches with superimposed orders and pediment, were already apparent in Middle Tennessee when architect Swope arrived; however, he attempted to correct and encourage the local exponents of such classical motifs so that their interpretation developed along with more understanding. His work here, fresh from Atlantic seaboard traditions, shows both his own ideas and his adaptation of the native styles. Back of Swope's classical ideas I suspect the influence of Shirley House at Charles City, Virginia, and the Bule Pringle Home at Charlestown, South Carolina, as well as the Fuller House and Drayton Hall, also in South Carolina. All of these display charming two-story porches with superimposed orders and pediment, flanked by low-hipped roofed wings. The twin chimney gable treatment is found all the way from the old Royall Mansion in Medford, Massachusetts, down the coastline to the Gilmer House in Savannah, Georgia; but, strange to say, they were never used in the central South except in Middle Tennessee.

The plan of the McGavock house is typical Middle Tennessee in germ arrangement, but is unusual in its use of entirely different verandas on opposite elevations. At the front is a small porch with superimposed columns; at the rear is a long veranda with square pillars extending through to the main cornice and supporting a second floor veranda. The entablature

" RANDAL McGAVOCK HOME "

is characterized by the use of cornice brackets and omission of the architrave. In considering these brackets as well as the frieze decorations and Ionic capital details, one is puzzled regarding the architect.

Maury County, Tennessee

RATTLE AND SNAP 1845

The Polks were builders. George Polk's house was the last built in Maury County by this famous clan who gave the United States its eleventh president—James K. Polk.

Such offences against orthodoxy are hardly ones to be committed by a designer fresh from the East. Was he so soon led to use a free hand in planning a frontier aristocrat's home, or were these details the work of a later restorer?

Carnton contains some excellent Georgian influenced woodwork and mantel pieces, but its chief glory, even in its prime, must have been the gardens. Randall McGavock, in common with nearly all his contemporaries, planned his home in the midst of elaborate grounds. The garden was designed by his wife, who consulted with her friend, Mrs. Andrew Jackson, exchanging bulbs, slips, and advice, with the result that the Carnton garden is almost identical with the one at the Hermitage.

Carnton was a famous stopping place for famous men. The drive eighteen miles over the dusty pike from Nashville was just long enough to encourage guests to pause for refreshment on the veranda with the McGavocks. President Jackson and President James K. Polk, who were at home in the county, Matthew Fontaine Maury and Felix Grundy and most of the other distinguished personages of the ante-bellum period were at one time or another recipients of the hospitality of this home. But its shadowed lawns and pillared porches have known a sadder hospitality than that of clinking glasses. When Hood made his desperate advance into Tennessee in November, 1864, and attacked Schofield's army at Franklin, the bloody fighting was done within gunshot of Carnton, whose residents remained to succor the dying and bury the dead. When the cold, raw dusk of evening fell, five Confederate generals lay dead on the great veranda of the house, proof of the terrific price which the men in gray had paid for their victory. Cleburne, Granbury, Adams, Gist, and Strahl—gallant guests who would never return. Their bodies were buried in the churchyard of St. John's at Ashwood, the home place of the fighting bishop of Middle Tennessee, Leonidas Polk. Colonel McGavock later donated two acres adjoining the family burying ground at Carnton as a resting place for hundreds of the Confederate dead. Today Carnton still stands, somewhat beaten by war and time, but sturdy and intact.

RATTLE AND SNAP — OAKWOOD HALL

Local pride, so general in the South, is easier to recognize than to define. It usually resolves itself, however, into a settled conviction that one's own locality is the best in the state, the nation, and—if one is pushed to it— in the world. Surely no community has cherished its local pride through the generations with more tenacity and charm than Maury County, Tennessee. After the Revolutionary War, most of the land in the county was included in grants to distinguished men who either came out themselves to establish their families, or else fostered settlements of immigrants from Maryland and the Carolinas. These various groups centered at Columbia and Spring Hill and Mount Pleasant. They brought considerable property with them, did the Pillows and Skipwiths and McQuires, and on the whole they prospered in the new country. Almost at once they began to build spacious houses for themselves and their sons and daughters who were intermarrying with their neighbors. For a century these homes have stood as symbols of agrarian economics, of family solidarity,

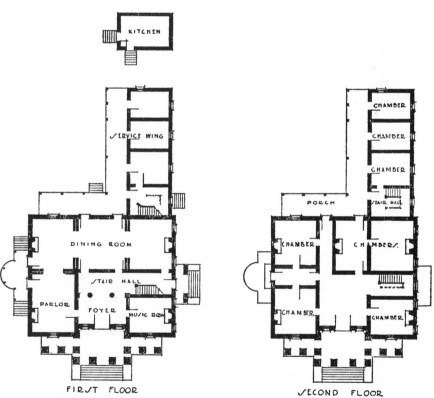

RATTLE AND SNAP PLANTATION

and of Maury County culture.

Probably the most famous clan in this section before the Civil War were the Polks. Shortly after the conclusion of the Revolution there came from North Carolina to Tennessee, Ezekial Polk and his son Samuel Polk, who were the grandfather and father of the man who was to be the eleventh President of the United States. Also with Ezekial came his cousin, Colonel William Polk, father of four sons, Leonidas, Lucius, George and Andrew. Samuel and his son, James K., were active in the political and social life of Nashville and other towns as well as Columbia; the family of Colonel William remained rooted in the soil of Maury County. One son, Leonidas, became interested in religion while at West Point and was ordained in the Episcopal Church. He became a Bishop, and assisted in the founding of Columbia Institute, one of the oldest schools for girls in the South, helping to design its faithful Gothic towers and to lay out its gardens. He also shared in forming the plans for the University of the South at Sewanee, with its serene and appropriate architecture.

The original Polk, Ezekial, later moved to Hardeman County and at Bolivar built his final home, Mecklenburg. Here he was always among the first citizens mentioned in West Tennessee History. He was destined to become the source of a great deal of bother to the loyal supporters of his grandson, James K. Polk, during his presidential campaign in 1844, because of his unusual philosophy, which he summed up in his epitaph on his tombstone:

Here lies the dust of old E. P.
One instance of mortality
Pennsylvania born, Car'lina bred
In Tennessee died on his bed
His youthful years he spent in pleasure
His latter days in gathering treasure;
From superstition liv'd quite free
And practiced strict morality,
To Holy cheats was never willing
To give one solitary shilling.

He can foresee, and for foreseeing
He equals most in being
That church and state will join their pow'r
And misery on this country show'r;
And Methodists with their camp bawling
Will be the cause of this down-falling
An era not destined to see
It waits for poor posterity
First fruits and tenths are odious things
And so are Bishops, Priests and Kings.

The "committee in charge" evidently got there before the photographer, because the reference to "Methodist" is today chiseled out.

The Polks were all builders. The Samuel Polks built a town house in Columbia, the county seat of Maury County, which was the boyhood home of James K. Polk, their illustrious son. Here young Polk started his unusual career, crowded with incidents portraying the Polk Clan characteristics. As a student at the University of North Carolina, young Polk walked six miles a day to meals rather than eat at the University commons. Beginning his university career as a studious recluse, he subsequently branched out and became extremely active in extra-curricular affairs. He was the only man to serve two consecutive terms as President of the Dialectic Literary Society at the University, and from the minutes of this venerable society, still alive, come many interesting little insights into his life.

Along one of the roads leading out of Columbia, the sons of Colonel William built their homes. Lucius in 1832 erected Hamilton Place, using skilled workmen which his father sent out from North Carolina. This fine old home is still occupied by his descendants, the Yeatman family. Three or four years later, Leonidas, soon to be made Missionary Bishop of the Southwest Territory, built Ashwood Hall, a more pretentious place, which he sold to his brother Andrew when he moved to his new diocese of Louisiana. Ashwood burned in 1874.

Because of its architectural interest, George Polk's house has been chosen here as the monument to the Clan. Rattle and Snap, built in 1845, is a glorified white-pillared house of the more glamorous type. An example of its grandeur is to be found in the ten great columns of the portico which were brought in sections from Cincinnati down the Ohio, up the Cumberland, and then by ox cart to Columbia. The labor on the house, however, was done by skilled local workmen, probably the same who had worked on the other homes of the family.

Many explanations are offered as to the origin of the odd name, Rattle and Snap. The most likely is that given by Mary Polk Branch, in her *Memories*, when she says, "My grandfather was playing a game of 'beans' with the Governor of North Carolina and some others. They played for 'script' issued to them as Revolutionary soldiers. My grandfather won the game, located the land, and named it for the game, 'Rattle and Snap.' It was in Middle Tennessee, then called 'The Territory of Franklin.'" The grandfather referred to is, of course, Colonel William Polk, who willed the piece of property to his son George at his death in 1824. The place is also known as Oakwood Hall, the name bestowed upon it by the Granbery family who owned and occupied it for a half century after 1867.

The plan of Rattle and Snap is a radical departure from that of the other local houses of the period. Gone is the usual great hall, extending the depth of the house, with its sweeping stair. In its stead one finds the minimum stair hall and open plan common to Louisiana. It is hardly fantastic to suggest that the Bishop's residence in Louisiana may account for this. The elevations have lost their traditional gabled roof lines in favor of the more Southern hip roof. The usual Middle Tennessee four columned pediment portico is flanked with the addition of three more columns on each side, which is very imposing. The order employed is good Corinthian, but the pediment is unusually flat. Entrances from the portico lead to a large foyer, from which open the principal rooms and the stair hall. The interior is elaborately done and agrees with the elevations in that the problem here called for the housing of a very proud and important family with a domain of almost feudal extent.

Built at a time when transportation was no longer a problem, the woodworking plants of the upper Ohio and the iron foundries of New Orleans furnished the necessary accessories. With local limestone, brick and timbers the builders could achieve strength and beauty. The excellent interior woodwork in the mantels, columns, cornices and the like indicates a very definite design, but here (as in other instances), no architect is recorded. The woodworking plants of Cincinnati, Louisville, and Pittsburgh are known to have furnished the plans and details for such items as windows, doorways, interior cornices, and fireplaces in houses similar to Rattle and Snap; and the general design could have been the result of tradition plus the culture of the builder.

On each end elevation is a lovely one-story covered entry. On the right it is a pure classical four-columned Corinthian order of wood and stone, while on the left it is a very delicate, but elaborate, porch and balcony cast iron assembly, projecting from the house.

The old house is still equipped with the early elaborate and unique system of service pull bells. On the service porch is a series of eight bells, ranging in size from two to six inches, all connected by small wires to the various rooms. Each servant knew his bell tone and, if he were a personal servant, he knew exactly where to repart.

The original gardens, which are said to have been very fine, have completely disappeared; but the approach, which is about a quarter of a mile long, with its beautiful setting of oaks, clearly indicates that the whole ensemble was in keeping with the importance of the house.

Other Maury County houses, showing definitely the same influence as Rattle and Snap, are the Mayes-Hutton Place in Columbia and Manor Hall and the Frierson Place at Mount Pleasant. The orders employed, the predominating corner pilasters, and the main entrance door details suggest that the same source of design prevailed in the whole community.

CLIFTON PLACE

CLIFTON PLACE is a living tradition. Many ante-bellum houses seem like museum pieces; to walk through their empty halls is to hear the past echoing like one's own footsteps. But here is a complete plantation with busy house servants and many farm hands to tend the broad acres just as it was a century and more ago. Continuously through the years, except for the interval when it was occupied by Federal troops during the War, the estate has known the round of sowing and reaping, of birth and maturity which marks a great stock farm in the bluegrass country. The pattern of self-sustaining agrarian life has never been broken.

General Gideon J. Pillow, leader in the United States Army during the Mexican War and subsequently a Major-General in the Confederate Army, built Clifton Place in 1832. Like the Polks, the Pillows were lavish builders, and this estate is only one of the several Pillow homes in Columbia. After the death of the General and Mrs. Pillow, the place passed to a son-in-law and then into the hands of Colonel J. W. S. Ridley, grandfather of the present owners. Thus maintained and cherished, Clifton Place is more than a monument to its builder; it is the embodiment of a civilization. One recalls the words of Ruskin in his *Seven Lamps of Architecture*, regarding men who are founders of homes.

. . . "Having spent their lives happily and honourably, they would be grieved, at the close of them, to think that the plan of their earthly abode, which had seen, and seemed almost to sympathize in all their honour, their gladness, or their suffering—that this, with all the record that it bare of them, and all the material things that they had loved and ruled over, and set the stamp of themselves upon—was to be swept away as soon as there was room made for them in the grave; that no respect was to be shown to it, no affection felt for it, no good to be drawn from it by their children; that though there was a monument in the church, there was no warm monument in the hearth and house to them; that all that they had ever treasured was despised, and the places that had sheltered and comforted them were dragged down to the dust. I say that a good man would fear this; and that, far more, a good son, a noble descendant, would fear doing this to his father's house. . . . When men do not love their hearths nor reverence their thresholds, it is a sign that they have dis-

Maury County, Tennessee   CLIFTON PLACE 1832

General Gideon J. Pillow built Tennessee's typical white-pillared house. It was one of three in Maury County built by the same family. The General was a friend of James K. Polk.

honoured both. . . . Our God is a household God as well as a heavenly one; He has an altar in every man's dwelling; let men look to it when they rend it lightly and pour out its ashes."

Any visitor enjoying the easy, generous hospitality of the Ridleys today can look about him in the ample parlors with their fine old mirrors and portraits and tell that here the "material things" which the builders "loved and ruled over, and set the stamp of themselves upon" have been preserved in honor.

On entering the house one is impressed with the very definite color scheme. The stair hall has wainscoting in deep buff over which is a wallpaper of cream background and deep buff and yellow medallion pattern. Woodwork is all white, while the all-over floor covering is plain medium green. Among the interesting pieces of furniture here are two Empire sofas, tall gilt French mirrors on marble top console and a grandfather clock.

The twin parlors have identical taupe velvet floor covering, with the walls in harmony. The first parlor boasts a theme in gold, gray and blue, while the second parlor reflects brown, mulberry and gold. Draperies are set in gilded canopies and are gathered to one side with tie backs of old-fashioned metal Greek acanthus in gold finish. The colors and materials of the first parlor are heavy brocaded Copenhagen blue, self-pattern, lined with cream and bound with wide silk braid to match. The second parlor draperies are mulberry and gold striped velvet, with a flounce at the top of solid mulberry edged with gold

Phyfe table and a colonial secretary.

The dining room color scheme is ivory, beige, rose and blue, with the walls papered in gray striped material broken with medallions of wild roses in a trailing pattern. The woodwork has a soft ivory finish and doors and mantlepiece are of wild cherry. The floor covering is a red, blue and cream Oriental rug. The draperies are of beige silk rep, tailored at the top with plain painted flowers bound in silk braid with metal medallion tie backs finished in gold. They hang from elegant gold cornices. The furniture is Empire mahogany. Notable among the accessories are milk-white, lace-edged bowls and a set of exquisite DuBarry china.

Clifton Place is typical of Middle Tennessee, with two stories and a cellar, and one story wings on either side. Outhouses, barns, and quarters, which are still standing in a good state of preservation, together with traces of the once formal gardens, enable one to reconstruct the life of the Cliftons.

Much of the material of the building is indigenous. The stone for the foundations was quarried on the site, and the brick was kilned on the place. For interior doors and trim, wild cherry was cut from the nearby forest, as were the hand-hewn pine timbers for structural purposes. The steps, column bases, and porch floors are of stone, while the column shafts, caps and cornices are of wood. The roof of the house is of tin, those of the outhouses of shakes. Cast iron balustrades frame the second floor under the entrance portico. The façades and the gable treatments with

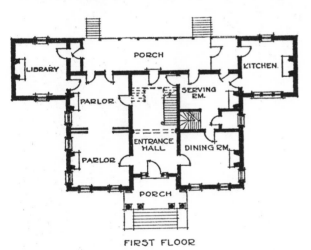

FIRST FLOOR

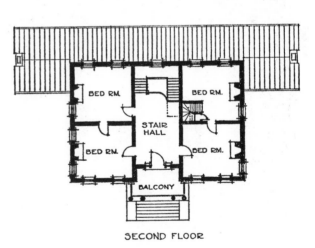

SECOND FLOOR

"CLIFTON PLACE"

braid; on all windows are glass curtains of round thread cream net with applique pattern of cream mull.

The furniture in the first parlor is principally early Victorian with occasional pieces of Duncan Phyfe and the customary French influence of the mirrors and marble top consoles. Furniture in the second parlor has a strong English influence with Chippendale occasional chairs featuring characteristic brass nail heads. There are other chairs of Adam influence, a Duncan

the stone buttress terminus of the cornice running the full length of the front and rear elevation are interesting. Twin chimneys are tied together with a brick parapet above the roof ridge line. The imposing, graceful entrance consists of a projecting portico two stories high, with four Ionic columns supporting a pediment of the same order. The entablature and pediment would be good classic stone proportions, but when executed in wood they are too heavy. They

Lauderdale County, Alabama

THE FORKS OF CYPRESS 1820

Still in the deep Tennessee Basin, but not Tennessee architecture, James Jackson's house has felt the influence of the Natchez Trace—hipped roofs, temple verandas, and cotton.

would be better lightened up somewhat by eliminating some members and simplifying others. At the second floor level a wood porch projects and is supported by the columns and enclosed by a cast iron rail between the columns. The entrance is through a large center hall in which the stairway is placed at the back and turns to form a landing beneath a window between the first and second floors.

THE FORKS OF CYPRESS

The Natchez Trace was by no means the only route of travel into the Middle South for the great tide of immigrants that poured west after the War of 1812. Indeed more settlers followed the rivers than went overland. Many thousands of them went up and down the great Tennessee River seeking the rich lands of which they had heard. As a glance at the map will show, the Tennessee rises in the mountains of East Tennessee, and loops down into the fertile lands of northern Alabama to Muscle Shoals (an impassable barrier for large boats until very recent times). Beyond that, it flows on, a safe and navigable stream, to its union with the Ohio in western Kentucky. Immigrants using this route either came down the river from East Tennessee to the good lands near the Shoals, or they journeyed down the Ohio from Pittsburgh and back up the Tennessee. The Trace, angling southwest from Columbia, crossed the Tennessee at Colbert's Ferry near Muscle Shoals. This junction of an overland route and a navigable river in a fertile region created in that section of Alabama another of those important centers of economic and social life that shaped the development of the far South.

There was a drift toward the good lands of northern Alabama along the Tennessee River as soon as the Indian threat was allayed somewhat; but the rush for holdings began with the development of cotton culture in the Old Southwest. By 1821 one-third of the cotton grown in America was reproduced west of Georgia; by 1834, two-thirds of it. Planters with capital and slaves came from Virginia and the Carolinas and Georgia to buy up small farms and new lands for great cotton plantations. Huge fortunes were made by the successful landowners, Lauderdale County alone producing about four million pounds of cotton in 1860.

Where economic growth is rapid, social advance of the people is likely to be accelerated. This section of Alabama, like other parts of the cotton kingdom, passed quickly through the phases of frontier hardships and poverty to attain a state of comfortable and even luxurious living. Historians have noted that the neighborhood from Huntsville down the river to the Shoals was settled from the first by gentlefolk and hardly "knew a rough regime." These planters, whom their enemies sometimes called the "cotton snobs,"

built handsome homes very early. The evolution of the pioneer home into the pillared mansion is not as clearly defined here as in the Nashville district. Rather we see "the grand manner" early in evidence.

To an architect who travels today along the few remnants of the shadowy old Trace where it cuts across the northern tip of Alabama, the significant thing will be his entry here into the cotton kingdom and the passing of the outpost of the horse country. He will see much in these ante-bellum homes that is common to Middle Tennessee and thus related to the older American traditions of Virginia and the Carolinas; but he will also note certain differences in plan, and plan for living.

Five miles from Florence in the junction of Big and Little Cypress Creeks, on a high knoll commanding this beautiful spot, James Jackson built, in 1820, his only home which he called The Forks of Cypress. The immediate estates surrounding it consisted of three thousand acres of land.

In plan it is a typical plantation home, Tennessee in character, except for the temple veranda façades which are in the Louisiana manner. There are center halls on both floors with the usual two rooms on each side. The chimneys, located on the outside walls, take up too much of that precious space and tend to crowd the windows. Farther south the planter who built on this pattern usually put the chimney pieces on the interior walls, reserving the exterior for light and ventilation. The façades, while traditionally Greek temple in form, are not constructed of materials usual in classical architecture. The pillars, instead of being of stone, are of wood, and the walls instead of brick and stone are of wood also. Fenestration and entrances show the influence of the Atlantic seaboard. The Ionic order is very good in the shaft proportions, but the capitals and entablatures have succumbed to local treatment, probably because of the limitations of slave labor and the failure of a good traveling craftsman to happen along at the proper time.

The outstanding feature of the Forks is its temple piazza. The temple type of house, a rare exception in Tennessee, was extensively used in Alabama, South Mississippi, and along the Mississippi River. This home, built rather early for the region and rather far north for the type is an outpost, as it were, of the architecture of the cotton kingdom—the first example we have considered of a way of building very popular in the Deep South.

James Jackson, native of Ireland, born of parents of little better than ordinary circumstances and well educated, came to America when he was a young man and settled in Nashville, Tennessee, during the time of Andrew Jackson. He came to Lauderdale County, Alabama, from Nashville in 1821, and eventually became a member of the Legislature where in 1830 he became President of the Senate. In 1817, with John Coffee and Andrew McKinley, he organized the Cypress Land Company which secured great tracts of land in the Tennessee River Valley from the Tennessee Land Company. This real estate concern opened up the lands around Florence in 1818 and sold off lots and country estates in that region. This section was excellent stock country for it was a recurrence of the Kentucky Bluegrass type and possessed clear, swift streams and good climate. Blue grass flourished winter and summer.

Following the domestic affairs of the Jacksons, one finds them a typically romantic family of the Old South. There were three daughters, Elizabeth McCullough and Mary Jackson, half-sisters, and May Ellan, a younger sister. With these young ladies in the offing, there was of course a great deal of gay life, a number of balls, and a continuous string of important guests. The Jacksons proved to be quite prolific, and from the 'twenties through the 'sixties, a continual train of children romped the hillsides of the Cypress estates. They were eventually to "carry on" for Alabama in the War-between-the-States. Mrs. Jackson continued as mistress of the Forks long after the War and during the Reconstruction days, and died at the age of ninety, leaving over one hundred descendants. James Jackson died in 1840.

Carrying on the tradition of early Tennessee influence and the Bluegrass regions, the Forks of Cypress became a very important live stock center. Its stables, much to the delight of Jackson, were second to none in the country. He stocked his stables with thoroughbreds, domestic and foreign, including the famous Leviathan from the stables of Lord Chesterfield of England, and Glencoe from the Duke of Grafton. Among the famous horses displayed and sired at the Forks were Magnolia, Gallopade, Iroquois and Peytona.

Surmounting the four slopes to the big house is said to have been four enormous orchards, two devoted exclusively to apples and two to peaches. In addition to this there was a double row of peach trees one mile long which lined the avenue leading to the quarters. Jackson is said to have kept only sixty slaves at the home place and the others in Mississippi on the plantations. In his quarters were extensive shops for both wood and iron work where all of the farm implements and wagons were made for his entire holdings.

On visiting The Forks today, one finds a very impressive sight. The big house still reposes on the highest crest of the hill commanding the countryside as though its importance was comparable to the Acropolis. Great white pillars depend on the deep shade of the veranda they border for their background, for here is missing the usual salmon colored brick wall of Tennessee, or the orange stucco of Louisiana. The outside chimneys are stuccoed and painted white; only the green blinds and the black shadows of the windows and the weathered wood roof break the spell.

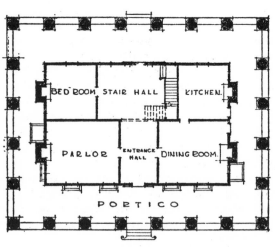

FIRST FLOOR PLAN

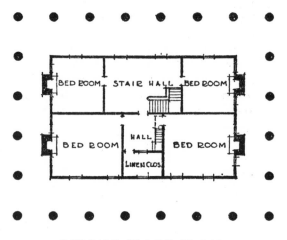

SECOND FLOOR PLAN

THE JACKSON HOME
THE FORKS OF CYPRESS —

Limestone County, Alabama

BELLE MINA 1820

In the deep Tennessee Basin, Thomas Bibb found limestone and blue grass. This is the outpost of the horse country.

BELLE MINA

BELLE Mina is a fine old North Alabama home in the rich Tennessee River basin, about forty miles above Muscle Shoals, and near the old town of Huntsville, in the edge of the cotton lands that lured settlers by the thousands just after 1812. It was built in 1820 by Governor Thomas Bibb; thus the time, the place, and the personality of its founder make Belle Mina a part of the early history of Alabama.

Thomas Bibb and his brother William, of a well known Georgia family, had a large share in determining the policies of the new region of Northern Alabama. William was the first and only Governor of the Territory from 1817 to 1819. The convention to frame the constitution of the state met in Huntsville, during July and August of 1819, at their very doorstep. William was elected Governor of the newly formed state, and Thomas, President of the Senate. Upon William's death in 1820, his brother succeeded him as the second governor of Alabama. Belle Mina must have been under construction during these exciting days. A distinguished procession of guests thronged its portals when Governor Thomas Bibb resided there, and during the years that followed.

The Bibb home is representative of many plantation houses in Alabama, although it shows little influence of Tennessee, or of Louisiana.

Belle Mina is totally devoid of any attempt at ornament or grand scale. Like The Forks, it has escaped the cornice brackets so favored in most of Alabama and Northeast Mississippi. As a whole, it is very pleasing in its simplicity. It really looks like a residence.

The plan is conventional with the usual central hall and flanking rooms. The great fireplaces occupy the outside walls in the Early Republic manner as found along the Atlantic seaboard. The front veranda (which by the way, is the only one), occupies one elevation and is unusually spacious and generous in it proportions, making it very livable. The service wing, one story in height, is at the left and includes such features as butler's pantry, kitchen, and rooms for two servants.

Belle Mina presents a distinguished type of white-pillared façade. Its main approach presenting the veranda is featured with six enormous Doric columns supporting the entablature from which springs the simple hipped roof. The order, in its entirety, is of the customary classic stone proportion, having escaped local liberties with the possible exception of the inter-columniation. Here we find a very interesting grouping with the columns placed three to the right and three to the left, leaving the central spacing wider to express the importance of the central hall and its impressive Georgian period doorway. The whole is altogether most expressive of plan and elevation—just another one of those amusing whims of the southern builder.

The walls are of brick, rich in earthen ranges as well as texture usually accounted for by the use of wood hand moulds, all of which furnishes an excellent background for the portico features of pure white. The floor of the wide veranda is paved with brick eight inches square, much the same color as the outside walls. The columns are of stuccoed brick and they have, I am informed, a core of 12" x 12" yellow heart poplar. The column capitals are of wood, as is the entire entablature.

The interiors are very simple, there being no elaborate cornices nor pilaster treatment. There are excellent large wood mantels with marble facings and brick hearths. The stairway is unusually fine in its impressive proportions as it makes one large semi-circular sweep to the second floor.

Mr. Bradley Bibb, eldest son of Captain Porter Bibb, is the present owner and lives there with his twin sister and very delightful family. Mr. Bibb is a gentleman of the Old South who is passing his last years amid the traditions of the three generations of Bibbs who have gone before him. Thomas Bibb died in 1839 and was succeeded by David Porter Bibb, his eldest son, who inherited Belle Mina and the entire estate and lived there during the Civil War.

David Porter Bibb was a very broad-minded man. While none of the neighbors disapproved of railroads, they all refused to permit right-of-way through their property. David Bibb notified the railroad authorities that they would be welcome to go through his property. Many interesting stories were told in this connection by the old slaves, one being that when the railroad was finished, the officials making the initial trip invited Mr. Bibb to ride to Huntsville with them. He refused, but said that he would order his favorite trotter, Old Boston, rigged up to the buggy and meet the gentlemen in Huntsville when they pulled in.

Mr. Bibb here held up the history he was telling me and related the story of the cyclone which struck Belle Mina on the night of July 16, 1875. He remembers the date very well because it was the date the second set of twins were born to his mother. It seems that the cyclone blew down the twin-chimneys to the front parlors at the same time these twins were being born. During the confusion, the cook had not reported for duty, and on the next morning it was found that she had given birth to twin boys. About the time this was all being "added up" and the strangeness thereof noted, the stable boy came running in with the news that a milk cow had given birth to twin calves.

The third owner and descendent was Captain Porter Bibb, Jr., who held the estate during the reconstruction days. Our host advises that Belle Mina Hall was the one remaining gay spot during those times, and assures us that it has lost none of its glamour. It did lose its formal five foot brick wall which so graciously encircled the immediate gardens and some several acres around the big house. Brick seemed to be plentiful with the Bibbs, and they gladly gave to their neighbors, for what would they do with enormous brick walls, when friends and acquaintances had no brick to build chimneys or fireplaces to keep themselves warm? Mr. Bibb chuckles at an old slave story concerning the burning of the brick for Belle Mina Hall. It seems that the kiln and clay pits were down by the creek, some half or three-quarters of a mile in front of the house. The old "Marster" had so many slaves they formed a line and passed hand to hand every brick from the kiln to the house site; therefore none of the brick was hauled as usual.

The railroad referred to was the old Memphis and Charleston Railroad, now the Southern Railroad. Not so many years ago, it was brought out in court that no right-of-way was ever dedicated to the original railroad; therefore the railroad could only claim the actual ground which their crossties covered, including the top of a cut and the bottom of a fill. Mr. Bradley Bibb has in recent years granted a power line a right-of-way through the property, but has sold them no land. Thus, though the Bibbs refuse to stand in the way of progress, Belle Mina Hall still remains the center of the estate.

PART TWO

MISSISSIPPI COUNTRY

THE Natchez Trace cuts across only the northwest tip of Alabama and then swings down into Mississippi through the fertile plateau at the headwaters of the Tombigbee. This great river system, draining Western Mississippi and Eastern Alabama, flows in a southerly direction to the Gulf of Mexico through the Black Belt, one of the richest cotton sections in all the South, comparable with the Mississippi Valley itself. At Mobile on the Gulf, the Tombigbee all but merges with the Alabama River, which comes down in a southwesterly direction cutting across the central cotton kingdom of the state. These two river valleys are virtually one section, economically and culturally, so much so that we shall include their contribution to white pillars in another chapter. However, there is no better example of the social and achitectural aspects of Central Alabama and Mississippi than the section around Columbus and Aberdeen, Mississippi, at the head of navigation on the Tombigbee. We may study it as representative of a wide region.

The history of the upper Tombigbee section repeats the motifs of the Southern frontier; Spanish or French or British explorers who built forts; American pioneers who drove off the Indians; small farmers who cleared the land; planters with capital who developed the country. Fort Choctaw, or Cedar Log Fort, was established at Old Plymouth near the present site of Columbus by the Spaniards, in 1790. In 1798, Mississippi Territory was annexed as part of the United States; but not until after a treaty with the Choctaws, did the country begin to fill with settlers. The first log house was built at Columbus in 1817 by Thomas Moore, and two years later, Robert Haden fetched the first stock of goods thence from Tuscaloosa. The Trace, as well as Jackson's military road built in 1818, and Robinson's Road down to the town of Jackson, all made Columbus a center of influence.

The town was located by its founders on a bluff overlooking the river, where the owners of the broad acres of the surrounding valleys built their homes. Crops were generally good, and the outlet to foreign markets easy by way of the river to Mobile. Consequently, the level of living at Columbus and thereabouts was high from early days. Fine houses were constructed and furnished with excellent articles brought from Mobile and from abroad. In 1821, Franklin Academy was founded, the first educational institution in Northern Mississippi. No wonder Columbus developed a well-poised and contented aristocracy which had little desire to alter greatly its pattern of life. Not until after the Civil War did the town permit a railroad to come within hearing.

Today, although it is a thriving town, Columbus retains much of its ante-bellum dignity and graciousness.

One who is privileged to come and go through the portals of its old homes is fortunate, because a fascinating problem in influences and styles lies before him. I have visited this section frequently for the express purpose of trying to determine whether the Trace or the Tombigbee—that is, whether the American Early Republic through Tennessee and Georgia, or the cosmopolitan French-Spanish civilization through Mobile—wielded the greater influence on the builders of Columbus. Unquestionably the best architecture here causes one to lean to the former theory, but there are unusual local influences strongly suggesting the presence of an architect with a yen for Gothic, whose attempt to create white-pillared houses from it is most interesting and worthy of note.

Because there is so much to be discussed about this vicinity, I will be forced to say a little about many houses instead of much about a few.

Columbus, Mississippi

THE HUMPHREYS PLACE 1840

Colonel McLawren found the best craftsmen in America, in plaster, marble, woodwork and masonry.
Brick here are so perfect that they are laid in one-quarter inch mortar joints without variation
detectable by a trained eye.

HUMPHREYS PLACE

SINCE leaving Belle Meade and Rattle and Snap, we have been visiting homes, simple in taste and indigenous to the place and the people, but here is pronounced evidence that we are to enjoy a home worthy of the description—elegant. In the Humphreys Place, the architecture, the examples of good furniture, draperies and equipment form a union of traditional and contemporary style in a gracious house of the past.

One of the two large scale mansions in and around Columbus, it was built in 1840 by a Colonel McLawren, a rich plantation owner, who, like many others in the section, had his great house in town. It is indeed on a much grander scale than its neighbors, the Street House and the Woodward House, not only in size but in the amount of space occupied by the gardens and estate. The illustrations given here shows the east street façade, which is duplicated on the west, or river front.

The plan is primarily of Tennessee influence, but very massive in scale, with sixteen foot ceilings and a monumental stair hall containing a fine staircase. Its detail and course of ascension to the third floor cupola are as beautiful as any I have seen in America. The interior plaster ornaments are very rich in cornice and ceiling. The detail of the window and door casings is of Greek revival influence, as are the mantels, which are exquisitely executed in Verde antique marble—an importation from Italy. The gilded window canopies, of a decided Louis influence, frame the equally graceful, flowing draperies.

The coloring throughout is low in key, as it should be to accentuate the rich medallions, cornices and the superb staircase.

The handsome furniture of the nineteenth century prevails throughout, displaying the best in pier mirrors, portraits and china. In this atmosphere, one falls under the spell of the ghost of a Strauss Waltz, a whisper of full skirts, a gleam of white shoulders!

The east and west façades are embellished by two-story porticos of four pillars each. The pillars are square, of stucco on brick with simple caps and base. The cornice is of Ionic influence. Wrought iron balconies over the entrance complete the ensemble. The other façades have a pilaster treatment of the same proportion as the portico pillars. The hipped roof is crowned by a glorified cupola having three windows on each side, making an observatory or light terminal for the great staircase. This conservatory is surrounded by a balcony or "Captain's Walk," which is bordered by a simple wrought iron rail. The balcony may be gained by any window since all of them extend to the floor. To the south are the customary service wing and kitchens. The garden plans were very effective, primarily because they were equally important from the river on the west and the street on the east.

Two features of the materials used in the Humphreys Place attract notice. The brick are remarkably good for homemade brick, being nearly mechanically perfect, with quarter-inch mortar joints. The roof is slate with wood gutters of hollow logs eight inches in diameter.

Columbus, Mississippi

THE OLD RICHARDS HOME, DATES FROM 1830

Early life and architecture in the upper Tombigbee Country is found here. Prototype? There were early inspirations, but none so picturesque.

OLD RICHARDS HOME

THERE are many old homes over the South exemplifying the sinister rustle of dead things but not the Old Richards Home. Among the earliest to be built in Columbus, Mississippi, it still reflects a full measure of the life and culture of its contemporaries while gracefully taking on the modern trends.

There is a sizable plot of ground between the iron entrance gate and the house, which is filled with native shrubs artistically developed. The iron fence across the front rests on a two foot brick wall and each span is tied together with brick posts. Except for the wrought iron, the entire fence is covered with ivy. The rest of the property is surrounded by a solid brick wall where American Pillar roses and ivy spill over the top in a most pleasing manner.

Immediately after the War, the house was bought by Colonel W. C. Richards and is now owned by Mr. J. P. Woodward, who has achieved an ideal of blending the gardens with the house. Large and small boxwood and an extravagant display of ivy is the theme of the grounds.

The house, in plan, is foreign to any custom of Tennessee or Louisiana, although in elevation it does lean decidedly to the latter type. There is the low-hipped roof line, the full ground floor of secondary rooms and the monumental steps, showing the importance of the second floor. The ground floor has not received a full story treatment, as the ground and the space under the porch is paved to carry out its function as a porch for the ground floor rooms.

The central porch section of the house is flanked by two wings which come out to the proch line. The porch is inclosed on three sides displaying the germ arrangement of the wing pavilion plan. The ground floor is of brick, and the main floor of wood with a wooden roof.

The Richards Home is one of the few early houses in the vicinity of Columbus with round pillars, fluted or otherwise. The columns are carved out of solid logs. While the order is good Ionic, its proportions and columniation are disappointing. The exterior stair leading to the porch, however, is a thing of rare beauty, entirely different from that of the Gorgas House in Tuscaloosa, or the Beauregard House in New Orleans, structures which are otherwise the probable inspiration of this.

THE STREET HOUSE
(The Colonnades)

ONE of the later white-pillared houses of Columbus, built in 1861 by William T. Baldwin, is now known as the Street House, home of Mr. and Mrs. Joe Street. In recent years, it has been restored under the direction of the author as architect. One plan and perspective illustrations show it before the restoration and enlargement. Another plan shows the additions and restorations. Here is a comprehensive study in the restoration of an ante-bellum house for modern living.

In original plan the Street House was foreign to any orthodox architectural tradition, although it was typical of many homes I have seen scattered over the South, especially the "T"-shaped, off-center hall plans of Mobile. On the first floor, the great stair hall was flanked by rooms on the south and on the north; but on the second floor, the room on the north was dropped and the plan thus thrown off center. Additions are shown in the alternate plan illustrated. The balcony of jigsawn wood over the main entrance is one of the most exquisite of the neighborhood.

The façade, shown in the illustration, is just as unconventional as the plan. The porch, six pillars wide and two deep, employs a square Mount Vernon type of wood order running through both floors and crowned by an entablature whose cornice projects three feet. And here we find again the cornice brackets which delighted the Southerner of this section. Somehow or other I find myself always coming to the defense of these brackets. In this case, they were removed when I undertook the restoration of the house, but I promptly had them replaced when the thirty-six inch projection of the cornice seemed to have a tendency to topple over. Here is a theory architects may have overlooked in their "reasons why." These houses, in most cases, did have heavy overhanging cornices which offered some protection to windows. For them to look too heavy would naturally have created some uneasiness. Optical illusion was corrected by adding a suport to the "shelf" and once again the builder was happy in a more secure feeling.

The façades have not been noticeably affected by the modernization, however in the interior an effort has been made to create modern living without offense to the traditional plan values. Bath, kitchen

Columbus, Mississippi THE STREET HOUSE 1860

Following the example of neighbors, Dr. William T. Baldwin built his home on the bluff of the Tombigbee and extended the garden down to the river's edge—a favorite building site adopted whenever a river was available.

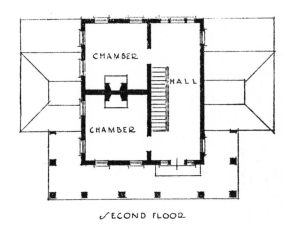

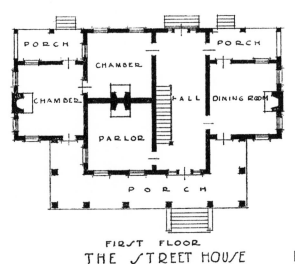

FIRST FLOOR

THE STREET HOUSE   BEFORE RESTORATION

and service, basement recreation room and terrace have been added, as well as complete new furniture and equipment—all with the theory of "sympathy with the Baldwin regime, but modern living for the Streets."

The parlor is a case in particular: the Streets having a flair for elegance wanted a "dressed up" parlor. It was determined to get away from the hopelessly standardized manner—the Victorian; so French influence with multi-pastels and violent color accent, so characteristic, was applied. The base of the color scheme is gold and blue with a trace of red and black mostly represented by the wallpaper and draperies. Natural finished pine floors of wide boards are cov-

ered by single rugs of the predominating blue. The marble mantel is indicative of Louis XV with an over-decoration of one elaborate plate glass mirror, reflecting for the most part, a single three-tier cut crystal chandelier which hangs from the center plaster cartouche in the ceiling.

The important pieces of furniture are done in gold leaf and petit point flower design tapestry, while other pieces are in dark teakwood, including table and mirror used as a wall piece.

Where our grandmothers were loath to risk their reputations by varying their manner of decoration, these restless moderns have a feeling for changes in fashion; consequently a blending of tastes gives opportunity for decorative variety.

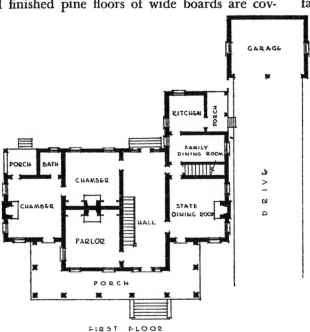

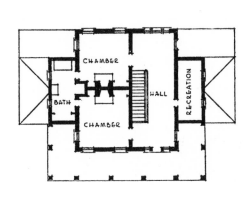

FIRST FLOOR   SECOND FLOOR

THE STREET HOUSE   AFTER RESTORATION

Columbus, Mississippi   THE FORT HOUSE 1850

Elias Fort established his plantation near the Alabama line, but built his "Big House" in town. Along with many others in Columbus, he was influenced to adopt Gothic architecture for white pillars. There are ten such houses in town.

THE ELIAS FORT HOUSE

Among the many wealthy immigrants to Mississippi during the Territorial Days were Mr. and Mrs. Elias Battle Fort and their three children, Mary, Martha and Elias B. of Nash County, North Carolina. Followed by a wagon train of slaves and household equipment, these North Carolinians came for the express purpose of reaping some of the promised benefits of the Cotton Kingdom. Here, near Columbus, Elias Fort established his plantation called Fort Springs, but built his mansion in town.

This town house was typical of the plantation "big houses" built in towns where "advantages" for a large and growing family of children could best be found. The estate occupied a whole block, about 500 or 600 feet square, or six acres, which rises from the streets on two sides. The house was the center of a small well-planned plot with a large oak grove on the slopes to the two streets. To the east was the formal garden of flowers and white shell paths bordered with box hedge. To the north were the well house, smoke house, carriage and stable groups, servant quarters and the vegetable gardens which included a small fruit orchard and berry patch.

Surrounded by a picket fence, such an establishment was calculated to be a family group, largely self-maintained, except for such heavy produce as milk and meats, which were brought daily from the plantation. Imagine a settlement of block after block of just such estates—all little establishments within themselves—and you have the ante-bellum town of Columbus.

In plan, the Fort House displays conventional Greek cross hallways with four rooms on each floor and verandas terminating two of the halls on the south and west sides. The elevation design, however, displays an interesting type and is unique and representative of several houses in and near Columbus, evidently designed by the same local architect. I say local, because in no other place in the South have I seen such adaptation of the Gothic architecture as is used here. The Southerner had used Gothic architecture as early as 1800 and as the years added culture to his civilization he grew in its knowledge through his travels. By the time Elias Fort built his house, about 1850, there had been a great number of Southern mansions built along Gothic lines. No doubt there was a sprinkling demand for this tradition, especially among the exponents of the "jigsaw" which was fast becoming the rage; but here we see an ingeneous and, by no means unsuccessful, attempt to conscript this old style for a white-pillared house. The French, in late flamboyant Gothic, were diversified and skillful in their use of stone tracery. No doubt had all of the leaded glass been removed from the side of some cathedral aisle, or a longitudinal section taken through the nave so as to display the aisle pier lines, we might find such an effect as is seen here.

The column shafts are octagonal, with well-formed bases and capitals of good Gothic moldings. The arches are correctly worked on four-point centers and the spandrels filled with lace-like tracery, all in wood. The balconies and porch baluster railings are strangely interwoven with patterns of kindred design. The entablature is of sympathetic design and proportion, with a simple dentil course at the cornice line. A low pitched metal covered roof keeps the height under control, which is fortunate, for the house reposes on top of a hill some distance from the streets on two sides. The ensemble from the street is a pleasing pattern of white lace which I must say is novel dress for the white-pillared houses.

Lowndes County, Mississippi

WAVERLY, THE ENTRANCE SALON 1858

Colonel George H. Young was from Georgia. He knew the fundamentals of ventilating engineering, and burned gas in his chandeliers. His "Big House" was the last word in luxury.

WAVERLY

One of the most interesting houses in the South and the glamour house of Columbus vicinity is Waverly, located on the west bank of the Tombigbee, about six miles from Columbus. It was built in 1858 by Colonel George H. Young, who migrated from Georgia about thirty years earlier, settled near Columbus, and became a large land owner. He was a man of stern qualities, a legal light, lord of his manor, and a power in the early development of the upper Tombigbee country. He reared his six sons and four daughters at Waverly amid such luxury as was considered extravagance, even in that extravagant period. The family enjoyed the labor of five hundred slaves—the house boasted bath tubs and gas lights.

The plan, conventionally simple in room arrangement, is featured by an enormous octagonal stair hall, fifty-two feet high and thirty-five feet across, with three tiers of balconies gained by three flights of free-standing wood staircases. This terminates in an octagonal cupola of the same size, serving as a fourth story lookout from which the entire plantation was visible. The four rooms flanking the hall are of equal size. The parlor contains ornamental plaster cornices and silk brocade wall coverings in a French "Louis" manner. The other rooms vary in taste from French to Egyptian, the latter in a second floor bedroom. The staircases, five in number, are simple and delicately done. Structurally cantilevered from the wall or free-standing in some instances, they

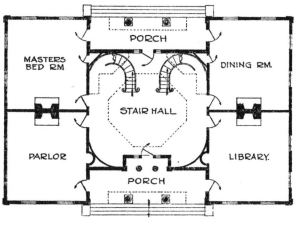

FIRST FLOOR

SECOND FLOOR

" THE WAVERLY PLACE"

Extracts from Colonel Young's diary reveal his many experiences with the imported skilled workmen whom he employed while building Waverly, and throw much light on building problems of the time. Family records show that Waverly was designed by an Italian architect by the name of Pond (which does not sound Latin). However, there was an architect in St. Louis at this time named Charles I. Pond, who possibly was the man. It is interesting to observe a custom of the day, that a man's name was always accompanied by his nationality—as though an added distinction, or means of recognition. The mantels and marble work were done by Richard Miller, a Scottish craftsman from Mobile; and the ornamental plaster was executed by two Irishmen from the same port. We have no record of the stair-builder, who was almost always a traveling workman. What a cosmopolitan and temperamental crowd of artists Colonel Young had on his hands during the months of construction. The rough frame and brick were wrought by his own slaves.

form an interesting treatment connecting the various balconies at floor levels.

The façades, north and south, are unlike any other of the Southern plantation types, although features from the Tennessee and Louisiana countries are both in evidence. The orders, Doric and Ionic, are worked in columns and pilasters to a pleasing combination in this unusual wing pavilion type. An extra refinement is added in marble steps and cast iron balconies. The main entrance doors have side lights of musical note with harp-shaped wood muntins.

In its heyday Waverly was a complete estate with numerous gardens, orchards, quarters, and a gas plant. All the chandeliers in the house are for gas, which was manufactured by burning pine knots in a retort set up on the plantation and operated by Colonel Young with slave labor. Too, information is available here confirming the use of bathrooms. The question often comes up in connection with life in these ante-bellum homes—why were there so few bathrooms? The answer is simple. When one needed the luxuries and necessities of a bathroom he had simply to command his personal servant, who usually was at his elbow during the day and slept on the floor just inside the door at night, and the bath was brought to his chamber. Hot and cold water in abundance were always ready in the service quarters.

Much of the comfort of life in such homes as Waverly depended upon the ministrations of these domestic servants, who were considered the aristocrats of the slave world. They were trained from childhood in the one talent of expert, obedient, courteous service. Their children, in turn, were taught the same arts and kept as household slaves if they showed any aptitude. The highest order among the house negroes was the personal maid or man servant; then came the cook, the laundress, the seamstress, the housemaid, the gardeners, the carriage drivers, the footman, and so on down to the stable boys.

The training and management of a large household of slave-servants was no small part of the responsibility of the owner of a plantation. The training for other purposes such as building trades was seldom attempted. Here, however, the slaves were evidently more adaptable to brickmaking, masonry, carpentry, and plastering, for it is of record that local planters developed these workmen through generations, just as they did domestic servants. A slave well versed in the mechanical trades was sometimes sold for three thousand dollars as against one thousand for a good field hand. Often one neighbor hired, or "loaned out" his skilled workers in exchange for similar favors when he in turn was ready to build. Many local good negro workmen today are descended from these excellent slave craftsmen.

THE MAGNOLIAS
(Of Aberdeen)

As a whole, the upper Tombigbee country presents us with a variety of architecture free from definite style tradition such as characterizes Maury County, or lower Louisiana, and not so glamorous either in scale or detail. At the crossroads of the South in the early nineteenth century, the Tombigbee builders absorbed many architectural ideas. They combined ideas of the Classical revival, the pediments and gabled roofs of Tennessee, the continuous temple piazza of lower Louisiana, the Georgia-Alabama custom of cornice brackets, the hipped roof and superimposed pillars of Louisiana's bayou region, and the tall graceful square shafts of the Maryland farmhouse. Locally the octagonal shafts and Gothic jigsawn spandrels seem to be the most original of all architectural elements. In plan, no two houses are alike and little tradition is recognized.

So typical of this feeling are the houses of Aberdeen, Mississippi, a neighbor town of Columbus and a little nearer the Trace traffic, although still on the Tombigbee. Aberdeen of course subscribed to the same life and culture as Columbus.

There is the Reuben Davis home, a dyed-in-the-wool variety of Greek Revival influence, with the Doric order supporting a gallery roof extending across the entire front and returning on each end to terminate into large rooms. There is the American Legion house with its Classical gallery of Ionic shafts arranged in pairs and predominating the entire front façade in three wide spans, supporting a nice wood coffered ceiling. Then there are a number of houses originally built by and still occupied by the Sykes Clan, from which we choose our representative house of upper Tombigbee Country life—The Magnolias.

The Magnolias estate was built in 1850 by William A. Sykes of materials and labor of local offering plus the skill and resources of Mobile, transported via the Tombigbee River. Through five generations of descendants, including the present owners, Dr. and Mrs. James Acker, Jr., this town house still holds its own. It was the good fortune of my office to plan restoration and additions in 1930.

The restoration of one of these old homes is a

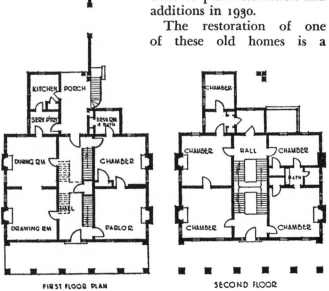

THE MAGNOLIAS

comparatively simple matter. It is always best accomplished by following the principle, "Do as little as possible." The original poplar timbers and hard clay bricks were sound and sturdy. It was necessary to add only twentieth century living requirements of modern mechanical equipment: baths, kitchen units, heating and insulating.

The floor plan is strictly conventional as far as the germ arrangement is concerned, but the stair hall is unusually interesting. I have found this parti in only one other house—the Cunningham Plantation House, near Cherokee in Colbert County, Alabama. One can see from the after-restoration floor plan illustrated, the convenience to say nothing of the decorative interest, in having such a stair where access to the second floor can be gained with equal rapidity and grace from either entrance — street front or garden front.

From the street the approach is through an avenue of dense, low-hanging magnolia trees which close off the house entirely except for an occasional white spot when the wind is blowing and shifting the leaves. Sun spots sift through the foliage to find a brick here and there on the broad aged brick walk partly covered with moss and ivy creeper. The façade, when close enough to be seen, presents six square pillars of Doric origin, springing from a first floor level seven feet off the ground and extending through the entire height to the roof line, creating a gallery of unusual charm and comfort. The other façades present white wood walls penetrated by well framed fenestration with dark green shutters.

Back to the stair hall, we find the keynote to the interior. In addition to the skillful treatment of this mahogany stairway, the room is made more alive with glowing color from the use of Bohemian glass overlights and sidelights at the entrance doorway. Here one finds the most logical position for colored glass; it is inviting and friendly and commands a position of importance. It seems to impart a feeling of peace and charm. The magnificence of the stairway spills from the hall above toward the opposite door, then divides in a graceful curve, and winds in opposite direction toward the front and back entrance on the other side of the hall. The rich crimson carpet picks up the glow from the glass and lends definite continuity to the space as it winds up the steps. The other side of the hall offers an interesting conversational group, making the hall seem of more importance than just a passageway.

Tall crystal banquet lamps add a light feeling to the atmosphere, as does the gray and white wallpaper in the Longfellow pattern. A large Victorian sofa with Empire influence, and numerous other interesting pieces give the desired anchor to this otherwise restless room.

The parlor is a room in which walls, floors, valances, and furniture reveal unrestrainedly the approved Victorian fashion.

A beautifully designed door leads from the hall to the garden porch, which is the same design as the front porch. Graceful steps with curves outlined in a wrought iron balustrade, lead to the garden and end parallel to a tall brick wall. This ivy-covered wall leads out to a separate dependency which was the original kitchen and connects that little whitewashed brick kitchen to the big house.

HOLLY SPRINGS
AND THE WALTER PLACE

In Mississippi by 1830 lands accessible to navigable streams had developed into thriving plantations and communities. The old Trace route had long since ceased to be of national importance and its bordering lands likewise had been occupied. Everywhere new land was in demand. Finally, the Indians, who had been continuously pushed farther inland, were all moved west of the Mississippi River and their old lands made available for settlement. Wagon roads and pikes penetrated into these backwoods communities, effecting communication to and from the steamboat routes so that finally (by 1840) the last frontier had been occupied.

Marshall County was one of these last frontier communities. Located northwest of Aberdeen, between the Mississippi and the headwaters of the Tombigbee River, Holly Springs—its county seat—was formally chartered in 1837. By 1850, the community was able to boast of the fact that it produced more cotton than any county within the state and was well on the road toward establishing a new cultural center in a fast-growing section of the South. Needless to say, white-pillared mansions sprang up almost overnight.

Most of the early Holly Springs' houses, as they now exist, date from 1850. At this time the Gothic and Classical revival influences were predominating in the entire South. In addition to these influences, the Jones McLlwain Iron Foundry, located in the community, had a decided influence on local design. This foundry produced balustrades, ornamental window lintels, column capitals, and bases, and Gothic tracery in such large quantities and ornateness of design that it seemed irresistible to local builders with plenty of money to spend.

The Walter Place, built in 1855, is the pride of Holly Springs community. It represents the larger

Holly Springs, Mississippi

THE WALTER PLACE 1855

Holly Springs' contribution to white-pillared houses dates from the middle of the nineteenth century. Here Colonel Harvey W. Walter was torn between the contemporary influences of Gothic towers, Classical revival portico, and a local flavor of cast iron detail.

and more glorified type of white-pillared houses, and is a good example because it embodies all of the architectural influences of the day in addition to making excellent use of materials at hand.

The street façades present a combination of these architectural influences in two enormous octagonal towers of Gothic which flank a central portico of Classic design. Cast iron is notable in the detail of both the Classic and Gothic. The order applied on the Classical portico is well-proportioned Corinthian with a light entablature which carries around all façades. The Gothic of the towers is of a slightly confused English interpretation, and its detail is inferior to the Classical motifs. All in all, the elevations are very interesting.

The plan (the one appearing here is after restora-

tion and additions) is more or less typical of Southern houses, but interestingly different in its use of the space within the octagonal towers. This feature is by no means unique in the South, as we find it a favorite among the Louisiana and French colonists, as well as many later builders who were guided by the spirit of the Italianate fashion. An example of the Louisiana variety is Melrose Plantation on Cane River. An example of the Italianate fashion is White Arches, the home of J. O. Banks, Columbus, Miss. The original plan consisted of the central hall flanked by two rooms and the tower on each side, both first and second floor. The additions to the Walter Place illustrate the adaptability of these old houses to modern living.

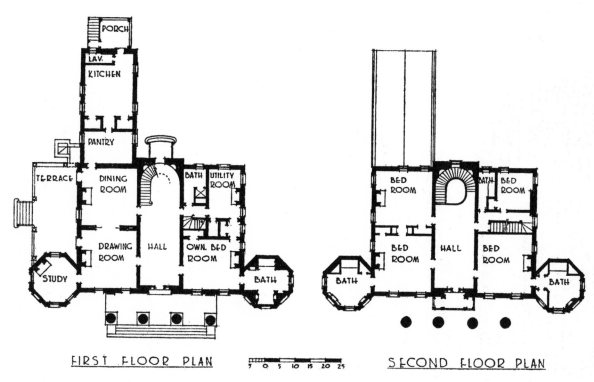

FIRST FLOOR PLAN   SECOND FLOOR PLAN

THE WALTER PLACE

Carroll County, Mississippi   MALMAISON 1845

Chief Greenwood Le Flore is one of the most unusual characters America has ever known. His house was designed by James Clark Harris, who became his son-in-law.

"We honor the memory of Greenwood Le Flore, not merely for the peace he brought to his community, not merely for the opportunities his diplomacy brought to the white race, not merely for his great energy and foresight that contributed so largely to the development of this country, but because he best represented the sturdy people of that day to whose untiring efforts and sacrifice may be attributed the later progress we now enjoy."

THESE words were recently spoken by an outstanding citizen at an auspicious occasion honoring the memory of Greenwood Le Flore, in the City of Greenwood, Le Flore County, Mississippi.

From Nashville to Natchez, the Trace traversed lands originally occupied by the great Indian tribes of the South. Of these, none was more civilized, peaceful and friendly toward the white men than the Choctaw Nation, which inhabited the Territory between the Mississippi River and the Tombigbee. The Choctaws traded and made treaties in turn with the French, the British, and the Americans, furnishing a regiment of warriors to Jackson in the War of 1812. Naturally, sometimes, they intermarried with the whites.

In 1792, a French Canadian, Louis Le Fleur, came up into Mississippi Territory, first to Franklin County and later to French camp in Choctaw County, where he established an inn on the Natchez Trace, and reared his large family by a Choctaw-French wife. In 1812, a prominent Tennessean, Major Donley, operator of a stage and mail line along the Trace, persuaded the Le Fleurs to entrust their twelve-year-old son, Greenwood, to him to be educated in Nashville. Five more years and Greenwood Le Flore, as he now spelled his name for convenience, went back to his home with one of the Donley girls as his bride, to become a leader of the Choctaws, who made him their chief when he was twenty-two. For over forty years, he ruled his people well, curbing their superstitions and avoiding quarrels with the whites. All of this time through his efforts, all negotiations and treaties between the U. S. Government and the Choctaw Nation were conducted. Time after time, he would make the trip overland to Washington in the interest of his Indian people. He became more and more persuaded that citizenship for his people, under the American flag, was an advancement over their savage customs, and finally that they should have a land all of their own west of the river. In '61, he steadfastly remained loyal to the Union, but gave shelter and food to all soldiers, whether in blue or gray. He died in 1865, the last chief of the Choctaws east of the Mississippi. Such a character is unparalleled in American annals; it is regretted that space here does not permit further discussion of his life.

Greenwood Le Flore built his first home, a log cabin, in Carroll County near a fine spring between the present towns of Carrollton and Greenwood. In 1835, he erected there a frame house, and not until 1845 did he build nearby his great house, which he named Malmaison to commemorate his admiration of the Empress Josephine. Here again on the frontier, we see a man advance from cabin to mansion in his own lifetime.

Greenwood Le Flore was in the prime of his life, recognized by his State and Nation, and surrounded by his large family, when he called in a young architect from Georgia (James Clark Harris) to draw the plans for a house. Timbers were chosen from his vast estate; every log was seasoned and trained workmen were engaged from abroad. Incidentally, during the long months which it took to complete the job, young Mr. Clark was married to Rebecca Le Flore, a daughter of the household.

Clark's plan is square, simple and conventional, with two halls forming a Greek cross, while a wing runs off at the rear forming an L for a state dining room sixty feet long. The elevations are each graced by a columned porch which extends to the roof line with a very elaborate wrought-iron balustraded balcony dividing superimposed doorways at the first and second floor hall terminals. The north and south portions have four square pillars, the east and west, two each. The roof is hipped, crowned by a double-deck cupola with a balustrade promenade all the way around.

Malmaison is a glorified example of the cornice bracket and jigsaw ornament. Here the main cornice is crowned by still two other cornices, one the promenade deck, the other the top of the cupola, forming a series of horizontal bracketed lines receding in height to the crown of the cupola. The mass resembles the decking of a great river steamer so much that it has been suggested that this was the inspiration for the jigsaw style. In the case of Malmaison, the elevations very probably were not the result of Le Flore's ideas, but of Clark's.

Le Flore gave his commission merchant *carte blanche* to buy the finest furniture in Paris, and there is a story that when the Duchess of Orléans saw the parlor suite, she inquired for whom it was being made. "For an Indian Chief in Mississippi," she was told. "What could a savage want with such furniture?" she exclaimed—and ordered it duplicated for herself. The more than thirty pieces of this furniture are of Louis XIV period, in gold leaf over hickory,

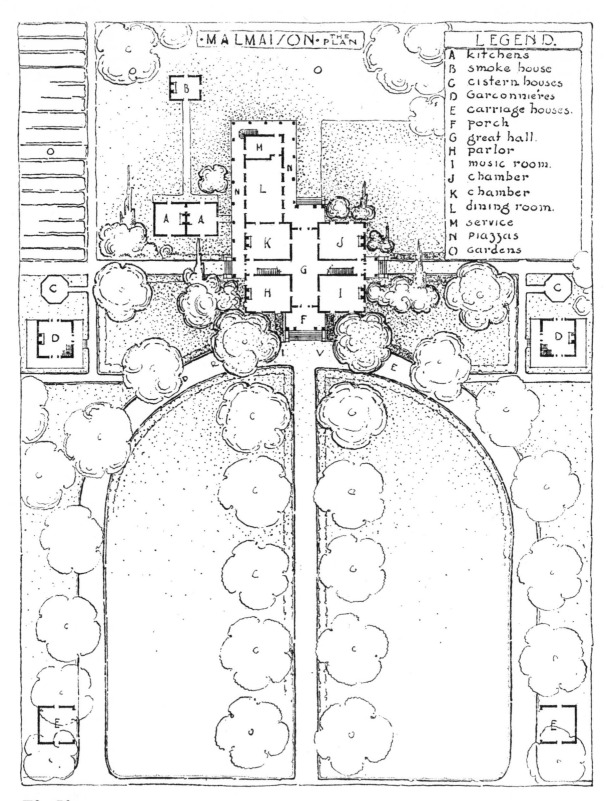

The following legend appears within the plan:

LEGEND.

A kitchens
B smoke house
C cistern houses
D Garconnieres
E carriage houses.
F porch
G great hall.
H parlor
I music room.
J chamber
K chamber
L dining room.
M service
N piazzas
O Gardens

The Plan

MALMAISON, CARROLL COUNTY, MISSISSIPPI

A Southerner never lost sight of the fact that he had plenty of space and that scale demanded he use it. Here we see a plantation group entirely covering a hilltop, overlooking an extensive valley.

upholstered in crimson silk brocaded damask. A table and cabinet of tortoise shell are reflected in two fine mirrors more than six feet high; while the marble mantel is surmounted by another magnificent mirror over seven feet in height. A clock and a pair of candelabra are of brass and ebony. On the floor is the especially woven Aubusson carpet with its rich crimson border and center design of roses, shading from light to deepest pink. The window valances are of gold leaf and the draperies of heavy silk damask. The shades are hand painted linen depicting views of Versailles, Fontainebleau, and other chateaux.

Malmaison is steeped with the personality and character of Greenwood Le Flore. The Indian fascination for the witchery and majesty of the primitive forest no doubt inspired him to place his home on the crest of a high hill, overlooking the forest country around. His love of, and loyalty to, the tradition of the French, he expressed in the house itself. There is every evidence of refinement in the many details of the architecture of Malmaison, from the silver escutcheon to the elaborate wrought iron balustraded balcony. The culture of Le Flore asserts itself again in his appreciation for the furnishings, reaching its height in the parlor, where every piece is a work of finished art. The sixty foot dinning room is evidence of his social proclivity. The eighteen foot table, with four extra leaves, could take care of the handsome silver and delicate china imported from France. There was enough china to serve one hundred people at the same time, and no two pieces of silver featured the same pattern. The dining room was connected with the kitchen by a narrow covered gallery fifty feet in length. Much of the cooking was prepared on the live coals of the great open fireplace. There were Dutch ovens with the tops hollowed out to hold burning coals—cakes and bread were baked in these.

At Malmaison we were offered some of the favored family recipes: The most popular of all Indian foods was *pishofa,* which is cracked corn, cooked until it is half done, then removed from the fire, dried, and mixed with equal portions of fresh meat, either pork or beef; then the mixture of meat and corn is cooked in a pot until it is done. This used to be the principal menu in ancient times at all the great gatherings. Another fine dish with the Choctaws was *ta fula,* made by parching corn, then sweetening it after grinding into powder. The most popular desserts were *wa husha* and *bahar.* The first was made of corn meal dumplings cooked with wild grape juice and sweetened with cane syrup. *Bahar* was made by mixing hickory nuts and walnut kernels, beaten into a pulp with parched corn meal flour, water being added until reaching the consistency of stiff dough. This was an exceptional delicacy.

The apprehension and anxiety for those engaged in the struggle of war caused Greenwood Le Flore to grow weary of a civilization lacking in faith with all he had been taught to respect and for which he had given up an Indian kingdom. His last request was for the stars and stripes to be held in the hands of his grandchildren, making a fitting canopy under which he breathed out his life; his fading vision resting until the end upon the flag he had loved.

Madison County, Mississippi

THE CHAPEL OF THE CROSS 1848

Margaret Johnstone's "God's Acre" still bears this shrine—a little brick plantation chapel and graveyard.

ANNANDALE AND THE CHAPEL OF THE CROSS

Southward into the heart of Mississippi, the Trace winds its way toward the bluffs on the Pearl River, where the city of Jackson now stands. To this vicinity John Johnstone, son of a former governor of North Carolina and descendant of the earls of Annandale, came in 1819, and called his homestead Annandale. He built a substantial log house, and brought out his wife and two daughters, Helen and Frances. Frances was married to a young North Carolinian, William Brittin, who settled in a neighboring plantation, Inglesides.

When John Johnstone died in 1848, he was buried in the garden of his home; but his widow, a devout Episcopalian, resolved to erect a church with a burying ground in his honor. And so she did, lavishing upon this little church in the woods the best that tradition and money could supply. The story of this Chapel of the Cross has been told to the author by Mrs. H. G. Thompson, one of the builder's descendants, in this wise:

"Sixteen miles from Jackson, on the old Jackson Livingston road, stands a little brick Church. One might easily pass it unnoticed, so hidden it is by great forest trees. It was built by Margaret Johnstone in 1848.

"The furniture was imported, as were the exquisite stained glass windows which mellowed the sun's rays on the glad day in May when the Chapel of the Cross was consecrated. There was a long line of Bishops and Priests, followed by all the neighbors from far and near, then by the faithful negroes, who had hewn the mighty beams, made the brick and labored so faithfully to build "De Church of de Lord for Ole Miss whar she gwine lay Ole Marster." His was the first grave in the little Cemetery at the east end. At the organ, little nine year old Margaret Britton played and led the singing, with her sweet childish treble. A long bench across the back of the Chapel accommodated the colored people whose mournful melodious voices joined with the rest in the familiar hymns taught by "Ole Miss."

"In God's Acre at the East end, under sweet magnolia shade, rest the blessed dead. One grave you'll find of a young lover shot down in his full strength, and brought to the little Chapel in dead of night. You can picture that solemn torch light funeral with the little sweetheart standing cold and still in the torches' glare, her sad eyes asking "why"?

"The grave of the young lover referred to here recalls the romance of the "Bride of Annandale," as Helen Johnstone is called in the stories of old days in Mississippi. How she fell in love with rich young Henry Vick, how he promised her he would never fight a duel, and how he was therefore killed by an unscrupulous opponent in a quarrel—all that is the background for the sad fate of the girl who remained faithful to him for the rest of a long and honored lifetime. His grave is in the churchyard of the little Chapel of the Cross."

But it would be a mistake to associate only sadness with these plantation chapels. They were greatly admired in their day as the center of neighborhood activities where joy predominated. A literary lady from the North, who visited a Southern plantation before the War describes it with enthusiasm:

"The situation of our chapel is romantic; and, being seen from all parts of the plantation, is an interesting feature of the scenery. It is about fifty-five feet long and built of stone; with turrets and mullioned Gothic windows of stained glass, and a floor of Tennessee marble. Its site is upon the verge of a green plantation, which overhangs the brook, and is, in its turn, overhung by a projecting spur of the Lion's Cliff. Majestic oaks embrace it, and ivy is trained up its walls. A broad lawn, crossed by graveled paths, surrounds it. These paths lead; one to the villa, one to the next plantation, and one to the African village where the slaves reside; for be it known to you that this beautiful chapel, the cost of which was $3,000.00 has been built for the slaves of the estate, too."

This same visitor also describes the Sunday when the Bishop baptized some of the pickanninies on the plantation:

"I have an amusing incident to relate of which our chapel was last Sunday the scene. The annual visitation of the Bishop being expected, the venerable lay-reader got ready some twenty adults to be confirmed, and forty children to be baptized. The Bishop duly arrived, accompanied by two clergymen. Our little chapel, you may be assured, felt quite honored with the presence of such distinguished visitors. There were several neighboring families present, who, with ours, quite filled the gallery.

When the time came to baptize them, the marble font being filled with fair water, the black babies were brought up by their ebony papas. The colonel stood sponsor for the boys, and his sister, an excellent and witty maiden lady, for the girls.

"What is his name?" asked a clergyman who was to baptize, taking in his arms a little inky

Madison County, Mississippi   ANNANDALE 1855

Margaret Johnstone spent a great deal of time in Tuscany. The Italian Renaissance is reflected in her home.

ball of ebony infancy with a pair of white, shining eyes. "Alexander, de Great, massa!" I saw a smile pass from face to face of the revered gentlemen in the chancel. The baby was duly baptized.

"What name?" he demanded of another Congo papa. "General Jackson, massa!" and by this name the little barbarian was duly made a Christian.

"What name?" — "Walter Scott!" "What name?" — "Peter Simple!" "What name?" — "Napoleon Bonaparte!" Splash went the water upon its face, and another ebony succeeded. His name was "Potiphar." Another's was "Pharoah." Another was christened "General Twiggs"; another "Polk and Dallas"; another "General Taylor"; indeed, every General in the American army was honored, while "Jupiter," "Mars," "Apollo Belvedere," and "Nicodemus," will give you a specimen of the rest of the names. The female infants received such names as "Queen Victoria," "Lady Morgan," "Lady Jane Grey," "Madam de Stael," "Zenobia," "Venus," "Juno," "Vesta," "Miss Martineau," "Fanny Wright," "Juliana Johnson," and "Coal Black Rose."

The water in the font, greasy and blackened by the process of baptizing so many black babies, had to be twice removed and replaced by fresh. The Bishop could scarcely keep his countenance as name after name was given, and the assistant clergyman twice had to leave the church, I verily believe, to prevent laughing in the church. The whole of this scandalous naming originated in the merry brain of the colonel's sister. Of course, the clergyman had to baptize by the name given, the whole scene was irresistible."

Some years after she had completed the Chapel of the Cross, Mrs. Johnstone began the plans for Annandale, a handsome home to take the place of the plain frontier home her husband had built long before. An architect from New Orleans was commissioned to design it, but the family has no record of his name. Whoever he was, he made Annandale an excellent example of the Italian Renaissance type of homes built in the middle of the century in the South.

There is a growing tendency among present day architects to call it Italianate fashion. Although this style did not satisfy the average planter's longing for tall columns, many Southerners did elect to model after the villas of Florence. Doubtless those who did so were influenced by foreign travel and protracted residence abroad. Many wealthy families in the cotton and sugar kingdoms spent only a few months of the year at their plantations, those when the weather was pleasant. The rest of the time they were in France or Italy. The "grand tour" was counted an essential part of a young man's education.

Annandale was far from conventional in plan. It had forty rooms, including dressing rooms—indispensible adjuncts of all chambers. The verandas rambled in and out, finally reaching the extremities of the three fronts. The exterior formed a delightful silhouette as it grew from one to two and then to three stories in height. The various projections were well placed, and the detail, although executed in wood instead of stone or stucco, was in most cases of good proportion. The orders, Corinthian and Doric, were all Vignola proportions, used as free standing columns and pilasters. The cornices, balustrades, quoins, window casings and arches were of wood.

Annandale was destroyed by fire about fifteen years ago.

Inglesides has all but crumbled for lack of care. Today descendants of this glorious clan of the Nineteenth Century reside together in a cottage nestled close to the Old Chapel, living in the past with all that is left of the "big houses", Annandale and Inglesides, among which are the following furniture and appointments, very worthy of note:

One Dining room table and twenty-four chairs all in bog-oak, richly carved Gothic.

Three silver service sets, serving twenty-four, for coffee, tea and water.

Seven pieces of French mahogany parlor furniture.

A marble clock with gold letters and hands flanked by Parian marble statues.

Set of Venetian glass finger bowls.

Full set of Gothic hall furniture, and many fine portraits of illustrious ancestors.

Claiborne County, Mississippi

WINDSOR 1861

The Cotton Kingdom has gone, along with Smith Daniell's Castle. Now the fields still lap at the base of the while pillars and moaning winds sigh through their majestic ranks—the darkies named it Windsor Castle.

WINDSOR

It was in 1924 that I first saw the Windsor ruins. I was driving along a narrow clay plantation road between wide fields of cotton when to my right at a curve in the road, rising above a cluster of trees, appeared twenty-four stately white columns. They were arranged in perfect order, each on a pedestal of its own and crowned with a bronze-colored Corinthian capital. A band of wrought iron lacework seemed to hold them together like a spider's web from twig to twig. The scale of the whole was appalling, even the trees seemed dwarfed beside them. I could hardly reach to the top of the pedestals.

On my way in search of Windsor, I had stopped at Port Gibson for other information. Several elderly gentlemen were seated on the porch of a general store, one of whom volunteered to guide me over to Bruinberg on the river. I had seen Bruinberg on old maps and read about it in early River yarns. Together with Windsor "Castle," it promised to be a treat even greater than I had anticipated.

My new-found friend prepared me for a real surprise. He had spent his seventy-six years in and around Port Gibson, and remembered the old house from early days. It was, he assured me, an exact duplicate of Windsor Castle in England; it had a fish pond on top of the roof, and there were solid bronze caps that weighed tons on each column. My hopes were just about to lift me from the misery of the clay road, when the sight of the columns burst on me like twenty-four giants in a story-book. I was almost glad I found no castle to destroy my illusion of white pillars.

Where the natives had concocted such incredible information, I must find out. My friend of the general store porch still stuck to his story and try as I might, I could not alter his narrative; so I photographed and sketched until lunch, delivered my guide back to Port Gibson and returned to do some private investigating. Negroes never fail to impart information which, if not always accurate, is interesting; and I hoped to find some original slave families in the neighborhood. I did. An old Mammy, who claimed to be a hundred and two years of age, remembered Master and Missus at the big house; she had been ladies' maid for the guests. She was not sure about this and that, but she did "speck" this room was here or there. Other niggers had to "tote" water but she turned a crank and her missus had bath water.

After an hour's conversation with her, I had reconstructed the plan and elevation of Windsor from her memories. Upon checking up with the ruins, I found that she was positively right. The plan was a Greek cross with service and dining ell off the right side of the house and the elevation like that of many West Feliciana Parish plantation homes. The column capitals were of New Orleans cast iron, along with the bases, veranda railings and lintels. The "fish pond" was the tank on top of the deck roof where water was pumped through lead pipes by darkies and heated by the sun for "milady's" bath.

The name of "Windsor" apparently was concocted by the negroes living near when they heard the eerie music the wind made through the huge capitals.

Windsor was built in 1861 by Smith C. Daniell. Its completion was, in a way, a climax of the race for supremacy in handsome mansions in that neighborhood; it was the pride of Bruinberg. The other great houses, begun with the intention of outdoing Daniell, were halted by the War. Their half complete foundations still remain, the brick having been used again for other purposes in later times. All of the architectural remains around Bruinberg substantiate my belief that these aristocrats did build with a good deal of envy and rivalry. Everywhere there seems to have been a desire to build a larger and finer house than the countryside had hitherto seen. Nor were the builders always old and white-bearded men who had taken a lifetime to accumulate their fortunes. Smith Daniell was thirty-four when this, his first house, was finished. Had his neighbors succeeded in building more impressive structures, I have little doubt that he would soon have given Windsor to his oldest son or daughter and set about re-establishing his prestige as owner of the finest home in the county.

The Spring of 1861, however, came as a smashing climax to such ambitions along the Mississippi. It was, perhaps, as well that Daniell did not live to see Grant's army camping on his plantation, in the campaign against Vicksburg. After the Battle of Port Gibson, the Union soldiers used the mansion for a hospital. A Dr. Loid, the chief surgeon, himself a gentleman of culture, is credited with saving the house from destruction. It survived the War and was destroyed by a fire of unknown origin in 1890.

As we leave Port Gibson and push on along the Natchez Trace, the shadowy old road is rich with historic memories. About forty miles above Natchez is the spot where Aaron Burr surrendered to General Mead near the haunts of his confederates. Blennerhasset, one of the most prominent of them, returned later to the county, bought a plantation which he named La Cache, and lived there for a number of years. A few more miles and we pass through Washington, a village that was once the capital of the state. Then, like the intrepid travelers of the early days, we near our journey's end. There are the bluffs along the river—and Natchez.

Natchez, Mississippi CONCORD FROM 1794

Here was Concord in all its glory. The flags of four nations have been raised over it since Don Gayoso, the Spaniard, built Concord. Now the masonry steps and iron baluster rails are all that are left.

CHAPTER IV

NATCHEZ AND THE FELICIANAS

CHAPTER IV
Natchez and the Felicianas
PART ONE

NATCHEZ

To understand the ante-bellum South, see the homes of Natchez. Here the past and the present make up a living tradition, unbroken and, in essentials, unchanged for a century and a half of colorful and significant living. When George Washington was President, the Spanish outpost of Natchez was set high where its hills overlook the yellow Mississippi and the wooded valleys across which might come Indian hunters or white men bearing foreign flags. Some of the buildings of that regime still stand. When Sumter was fired on, these hills were crowned with the white-pillared homes of rich planters whose hospitality was memorable. Crushed by Grant's blue armies that were more inexorable than the tawny Mississippi, the planters were defeated; but their houses mostly remain. Nowadays Natchez is a modern town of fifteen thousand or more; modern, that is, in highways and conveniences and *savoir faire*. In most of the old homes people are living charmingly in a twentieth century manner. Yet the pattern of life in Natchez is distinctive just as these homes are distinctive. No apartment houses and mail order catalogues have as yet revolutionized the habits and tastes of its citizens. The pages of southern history are legible and clear in Natchez.

Happily a visit to Natchez in these times is easy to achieve. Not everyone, to be sure, will be lucky enough to be a guest in the homes where Lafayette and Henry Clay were once entertained; but he can go in the springtime during Pilgrimage Week when the ante-bellum homes are opened to the public and see them at their best. The roads are good and the weather usually pleasant along in March when the first families of the town play host to increasing crowds of visitors. The moderate fees for sightseeing are applied toward the restoration of old homes.

"But why Natchez?" a stranger may ask. "What can there be in some old houses in a river town to explain the culture of the South?" And the answer is not easy to put into words. The civilization of the South, before and after the War, is much talked of, much portrayed in fiction and on the screen; but it is very elusive. It is more easily experienced than explained.

Too often the idea prevails that life a century ago as we have been describing it along the Trace in Natchez and Columbus and Florence and Columbia was narrow, dull, sectional, because these towns were numerically small and geographically remote from the Atlantic seaboard. As a matter of fact, the privileged classes in such villages at that time were more cosmopolitan than the corresponding classes in most American cities today, and this in spite of our automobiles and radios and newspapers. In the first place, their settlers came from many different states and countries, bringing with them the backgrounds and customs of their former homes. Moreover, communication with the outside world, while it was very, very slow, was easy along the rivers, and people of means traveled rather frequently. These families were connected by blood and friendship with similar groups in many other towns in the South and East; and they maintained such ties rather more systematically than is the custom today. Contacts with Europe were also usual, both for business reasons and for pleasure. The result was a high degree of individuality in manners, tastes, and possessions among the early Southerners.

At the same time these plantation owners and their neighbors developed certain traits in common, chief of which was the agrarian habit of building and keeping a homeplace. Patriarchal in their family life as in their economic political relations, they invested lavishly in their houses and in furniture and equipment for them. These establishments became the centers for the social life and established the cultural pattern of the community. Nations and regions express their culture in various ways—in public buildings, monuments and roads, in factories and business offices, in colleges and churches. The leaders of the Old South in towns like Natchez expressed themselves most fully in their homes.

Another common trait among these ante-bellum Southerners was hospitality, which they practiced and developed even beyond the habit of the rest of the American frontier. Food was plentiful; houses were spacious and easily heated; in families already numbering twelve or fourteen members plus half a dozen relatives, no visitor ever felt himself an intruder. At Natchez hospitality became a code, a ritual of exquisite courtesy. Apparently, every famous traveler of the early nineteenth century made it a point to go there and be entertained in the homes of the Upper Town. Today, as the Pilgrimage Week sightseer goes his rounds, the boast of many a guide is that "Henry Clay was an honored guest here and spoke of it afterward in one of his speeches in the United States Congress," or "There is the piano on which the accompaniment was played for Jenny Lind when she sang in Natchez in '51."

This unity in diversity, this combination of a regional pattern of life with personal individualism is, of course, not unique in Natchez or the South. It is American. Especially is this true of the emphasis on the home as the central unit of society. His home as the crown of his life work was the goal of every successful man in the earlier days. It was the fruit of his endeavor, the symbol of his position in the community. He built it for himself and for his children, confident that they would want to live in it. Old-fashioned and naive this may seen to a generation accustomed to city apartments and the methods of the realtors; yet perhaps the America of tomorrow will reassert the importance of home building. The old fashion may become the new fashion.

Much can be learned at Natchez about the conditions under which building was carried on. The records there are fairly complete although often traditional; and the evidence as to how the work was done is so continuous that we can fill in the gap left at the houses along the Trace. The skill and inventiveness of the owners, which we must surmise often-

times at other places, is revealed fully among the educated and inventive builders at Natchez, who often enough remind us of that greatest of Southern gentleman-architects, Thomas Jefferson. Thus at Landsdowne, built in 1852 by George M. Marshall, there was a privately operated coal-gas plant. Everywhere the utmost use was made of native materials such as stone, timber, and brick made on the place. At Homewood, for instance, the timbers were seasoned for eighteen months, and the formula for mortar was tested for eight months before the structure was begun. At the same time, no false patriotism limited these men to local products alone. They placed their orders for grill work in Spain, stained glass in Belgium, chairs in Paris, and paintings in Italy. Cosmopolitan in their tastes, they counted nothing too good for a mansion overlooking the Mississippi. Another interesting chapter in Southern building is to be read in the several homes that were left unfinished when the call to arms came in 1861. At Longwood the workmen left so suddenly that they dropped tools and paintbrushes on the floor—which are still there. Some sense of history or of their own dignity has happily prompted the citizens of Natchez not to destroy their own past.

Indeed there is an embarassment of riches at Natchez for visitor or student. Whole books have been written about its history and architecture, and with good reason. To have to make the choice of half dozen homes to discuss in this chapter is a grave responsibility and one I would gladly avoid. I can only select those which, to the eye and mind of an architect, seem most significant because of the trends they reveal and because of their beauty.

CONCORD

When the Natchez Country was a province of Spain from 1779 to 1798, the representative of His Majesty was a charming diplomatic gentleman, Don Gayoso de Lemos, to whose intelligence may be attributed the mild rule and peaceful atmosphere of that twenty years. He had the good sense to welcome Americans into the territory, and to foster friendly relations along the river in order to build up the trade of his settlement. As a result his home became the center of a rather brilliant provincial society.

Don Gayoso built Concord in 1794 for his own home, naturally proposing to make it reflect the dignity of his position and his nation. He furnished it in Spanish style with iron work and oriental hangings. Tradition insists that he maintained a retinue of uniformed servants. After Natchez passed into

American hands, the house was occupied by Governor Winthrop Sargent until 1808 when it was sold to Don Estavan Minor, a Pennsylvanian who had served as an official under Spanish rule. Minor married into the Lintot family and so was connected with the ill-fated Philip Nolan, whose career inspired "The Man Without a Country."

For more than a century Concord perpetuated the formal, rather lavish hospitality of the Spanish aristocracy in early Natchez. In it we recognize the prototype of many great houses in the community. The elongation and detail of the orders applied here, along with the pediment, are found later in Gloucester, Arlington, and Rosalie. In those houses, in spite of the composition of mass and detail that show classical and Georgian periods of English influence, the general outlines of the early Spanish mansion of the West Indies reveal themselves.

Concord was destroyed by fire in 1901. The illustration here is copied from an old print showing the house in its prime. I have tried in vain to secure a plan of the interior. This much, however, we are told. The ground floor which housed the domestic animals as well as the carriages and some of the servants was substantially built of brick with a simple division of rooms formed by the columns and upper verandas, opening onto the lower shelters. The principal rooms of the house were on the second floor and opened onto the galleries which encircled them. Some chambers were in the third floor attic space. This arrangement is, of course, suited to the Spanish manner of life and may be seen in the West Indies, Mobile, and elsewhere. We have no trace of the garden, but it surely was commensurate with the house.

We have discussed both the superimposed and single orders employed in many of the great houses of the Natchez Trace and Nashville. We have even had examples, such as the Randall McGavock house in Franklin, Tennessee, where the single orders were employed on one elevation and superimposed orders on another; but here in Concord we find a great, single column extending from the ground to the second floor roof line, alternating with a light balcony post, which by proportion and function is virtually a superimposed column.

In elevation Concord is probably the best early example of the Southerner's ingenuity in using classical orders to provide the much needed veranda. As we have observed in many examples, these early builders recognized no predetermined mathematical proportion in the orders, and if they had bothered to master such theory, they would doubtless have lost sight of it in practical application. There is nowhere to be found any prototype for the façades of Concord, and we may assume (with confidence) that there was no architect (as we know him today) to labor over its lines. There is much speculation as to the addition of the white pillars—some claiming they were original and pre-date the Classical Revival Period.* In 1803, C. C. Robin, a French traveler, said of the lower Mississippi River houses, "Some of the houses are built of brick with columns . . . " That columns did grace the porches of houses in the Deep South, many years before the recognized so-called Classic Revival Period began, is a fact established by many such contemporary writers. Others differ. At any rate such homogeneous structures as Concord are the result of an uncanny skill in composing a mass, plus a perfect understanding of the elements of architecture as applied to the needs of those who were to live in the house.

A detailed examination of the façades shows that many great pillars, each springing from the soil on its own base, extended through two floors and supported a light wood pediment and the hipped roof. The pediment formed a shelter for the monumental twin staircase, bordered with wrought iron balustrades meeting at the second floor on a balcony, and all enfolding a driveway through which carriages disappeared after depositing their passengers. Having attained the second floor by means of this great welcoming stair, one wandered around the entire perimeter of the house, entering any room through spacious casement openings which served as windows and doors. The French and Spanish builders delighted to magnify the importance of such stairways.

Although the main or second floor veranda extended around the house, it rested partially on brick walls of the lower story and, at intervals, on the columns. It was carried by the four groups of pillars in two different compositions, the pediment and the hipped roof, thus shifting the burden from one to the other, all without any seeming uneasiness in appearance. There seems, all in all, to be no more original or dignified solution of the problem in all the Southland.

(* Supposedly 1830)

GLOUCESTER

ON the old Trace near Natchez stands Gloucester, a fine mansion that has been admired by travelers since 1800. In 1808 Fortescue Cuming, as we have already noted, spoke of it as a "handsome brick house" denoting "more taste and convenience" than any he had seen in the Territory. In 1831, Ingraham echoed the praise that it was "a handsome mansion," and commented upon the fine trees which still surround it. Its present owners welcome thousands of visitors every Spring.

Historic as well as beautiful is this old home. Most authorities agree that it was built in 1800 by David Williams, or his family. A member of the family married Maria McIntosh, who later became the wife of Winthrop Sargent, first governor of Mississippi Territory when it became a part of the United States in 1789. Governor Sargent, as his name would indicate, came of an old aristocratic, strait-laced New England family; and when he took up his residence at Concord, the home of the former Spanish governor, he found much to disturb him in the cosmopolitan ways of the household and the town. After his term of office had expired in 1808, the governor purchased Gloucester for his permanent home, and it seems certain that he became more reconciled to Mississippi life under its roof. He developed the plantation and gardens, furnished the house lavishly, and when he died in 1818, directed that he be buried in the willow yard on the place. At least one well-informed writer on Natchez homes believes that the stairways at Gloucester resemble those of the old Sargent home in Gloucester, Massachusetts; and it may be that the governor copied details from his ancestral place in developing and remodelling his Mississippi estate. In all events, the story of Winthrop Sargent is the evolution of a strict, somewhat puritanical, territorial governor into a Southern planter whose remarkable will discusses at length the problems of his landed estate. The one surviving son of the Governor inherited the place and during the War was brutally shot by Union soldiers as he hospitably opened his door to them.

From the architectural point of view, Gloucester is not only beautiful but significant. Here a plan typically Louisiana in arrangement, is combined with the later pediment portico and the great pillars of the Classic orders. The plan in accord with Bayou Country tradition is the three rooms wide and one deep type, with verandas on the two long sides onto which the rooms opened.* Like most Southern houses, however, it has enough variation to create special interest. The octagonal ends suggest a like-

---

* See Chapter VI for Louisiana plan types.

ness to Woodlands and The Laurels in Philadelphia. The use of inside wall space for the fireplaces is good Louisiana practice, but not the placing of the stairways on the interior instead of the verandas. Gloucester moreover has original features. There is a large

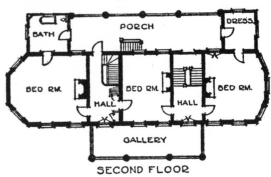

SECOND FLOOR

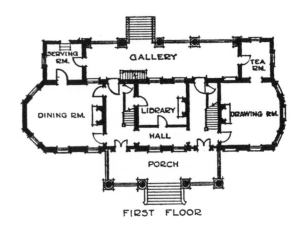

FIRST FLOOR

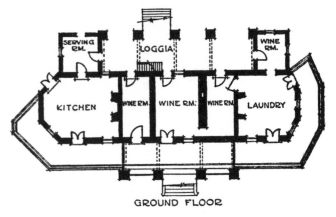

GROUND FLOOR

basement with heavy barred windows and doors, which must have been intended for definite uses. In the plan drawn here, I have recorded reasonable conjectures and local opinions as to the uses to which the space was put. This basement is surrounded by a dry moat.

The first and second floor plans are, even when judged by modern standards, very convenient and well arranged. They could well be imitated by home builders today. The twin stair halls, terminated by beautiful fanlighted doorways, contain graceful staircases with mahogany trimmings. These halls are connected at the north portico with another hall which gives access to the library. The octagonal drawing room at one end is matched by the octagonal dining room at the other, the whole effect being one of dignity and spaciousness.

Elaborately furnished in the early days, Gloucester has fortunately never been shorn of its treasures. Today its fine marble mantels and crystal chandeliers are intact. Priceless paintings hang on the walls, which include works by Canaletto (1697-1768), Salvator Roas (1615-1673), Jose Ribera "Spagnoletto" (1588-1656), Francesco Bassi (1642-1700) and other noted artists. The mirrors and books of past generations are preserved by the present owners, who have also added to the riches of the rooms.

Queen Anne furniture holds the spotlight in the drawing room, Bristol vases lend it color. The charm of the home is further enhanced by mantels of black African and Italian marble surmounted by mirrors. Furniture in the library is Empire, made of mahogany and trimmed with gold ornaments, characteristic of this period. An Adam rug in neutral shades covers the floor. This elegant and tempting room has many first editions. The music room boasts an Aubusson rug. Francois Seignouret of New Orleans made the Empire dining room chairs of mahogany, in the early nineteenth century. On the mahogany Chippendale china cabinet, inlaid with fruit-wood, is a grouping of elaborately designed Meissen ware.

Like its companion houses in Natchez, Arlington and Rosalie, Gloucester has a pediment portico on one elevation and a hipped roof veranda on the opposite one. The elevation from the north presents a rather formal pediment supported by four graceful Doric columns and full entablature. The fanlight entrances into the stair halls are correctly placed between the outside columns, and thus give the elevation a feeling of balance. The south elevation displays the conventional Louisiana colonnade, the columns, each on its own foundation, supporting a simple, light wooden hipped roof with a wood balustraded gallery at the second floor.

ARLINGTON

ARLINGTON is set back in a wooded park on the outskirts of Natchez. The mansion has been fortunate in its masters (and who is there to say that the masters have not been equally fortunate in their home?), because the traditions and spirit of ante-bellum life have been conscientiously preserved.

Arlington dates back to 1816, when it was planned and built by Mrs. Jane White, daughter of Pierre Surget, one of the pioneer French settlers in the Natchez country. It is believed that she engaged an architect-contractor from Philadelphia to design her home and superintend its erection. If this is true, she probably schooled him for several months in the Natchez tradition before she allowed him even to sketch; and then, as the results prove, he set about applying his knowledge of correct architectural principles to the problem of building Mrs. White a house suited to Natchez. That he had skill of an unusual quality is evident.

Arlington records the beginning of influences different from the French and Spanish elements in Natchez, destined eventually to turn the tide of building fashion away from Concord and its manner. Here, with the exception of the hipped-roof veranda of the rear elevation and the orders, there appears to be a definite abandonment of the early parent influences. The order employed is still Doric. The shafts are a perfection of the type initiated at Concord and used later at Gloucester. In the entablatures and low-hipped roofs flanking the pediment portico, we see other elements that can be traced back to the earlier less perfected ones at Concord. Here however the pediment, instead of springing from the farthest projection of the cornice as is conventional, springs from a point directly over the face of the frieze. This tends to strengthen the whole; for should the pediment spring from the end of so light a cornice, it would create an uneasiness and have a tendency to break off the cornice member of the entablature. Another new note is the Georgian detail in the entrance doors and interior work, a refinement hitherto lacking.

The plan is the old familiar one followed in Tennessee—the great hall flanked by two rooms on each side, upstairs and downstairs, with porches on the two broad elevations. One exception is the fireplaces, which are drawn in from the outside walls. The yard kitchens, storage, and service units are present. Here, apparently, Natchez for the first time has almost entirely dropped the French or Louisiana plan and adopted the Anglo-American one. In façade and in plan alike, this house shows that Natchez, in 1816, was arriving at a perfection in the white-pillared house that other communities did not attain until

Natchez, Mississippi ARLINGTON 1816

The white pillars of Natchez came a decade earlier than elsewhere in the Deep South. Jane White chose the conventional plan of the Tennessee country, in preference to that of Louisiana. Arlington, although built at a very early date, was indicative of the full development of the Classic period influence in the South.

almost 1836, and most of them not at all. By 1816, the early Louisiana style house had developed fully its typical mass and fan transoms; and Tennesseans had begun to add white pillars to their frontier plantation homes. By 1820, Natchez builders, beginning their use of pillars where Louisiana practice left off, achieved their greatest work.

Now to the interior. A description of all the treasures of Arlington would fill a book. So I am going to draw the reader's attention to the gold drawing room and say of the rest of the house, "Its furnishings are in keeping." This room derives its name from the upholstery and hangings of sunlight satin damask, and shows an orderly and symmetrical disposition of furnishings and decoration. The original French hand-blocked wall paper has been faithfully reproduced. Six candle sconces of chiseled ormolu done in graceful flowing designs are hung in symmetrical positions on the walls. They are of the suspended bracket variety disporting five candles in each grouping. The crystal prisms were very much in favor because of the brilliance and manifold reflections of the rays of candle light.

The rococo pierced-shell motif of the hand carved rosewood furniture repeats itself in the handsome gold window cornices. We should not be far amiss in saying that this room is done in the Adam manner. The period of greatest success of the brothers Adam was contemporary with the Chippendale period, and one sees in these influences of Chippendale a direct reflection of French and English.

There are seven sofas counting the three beautifully designed window seats upholstered in the same original brocade. The largest sofa is called the chaperone sofa, and a smaller one the engaged sofa. In Chippendale's day, the larger was called "Darby and Joan," oftentimes known as bar-backed sofas. The French term, "confidence," was frequently applied to the smaller of these settees.

The century old window draperies are of silk, five and one-half yards long, and the glass curtains are of handmade Brussells lace. Tie backs are handsome clumps of bronze grape leaves with bunches of white glass grapes spilling in a naturalistic manner. The table occupying the center of this room is of very rare leopard's wood from Africa.

Signed bronzes done by well-known sculptors rest on richly carved pedestals which were brought from the ruins of Pompeii. A gorgeous Chippendale mirror, made especially for the particular wall space which it covers, reaches from floor to ceiling; a pair of Chinese hawthorn vases in blue and white flank the mirror. The Aubusson carpet is the last word in making of the decorations a unified whole. As I have stated, gold prevails in this room and sparkles against the gray tones of wall paper. One might call this rich glow imprisoned sunlight. It is truly an excellent example of a master decorator house.

Arlington and Rosalie, to me, represent Natchez architecture at its best. No houses in America have had such success in adapting the classical elements in a domestic design. Their slender, graceful column shafts and exquisite light entablatures are totally void of any suggestion of Roman-Grecian classics proportions or even the governmental style of public buildings. They are just homelike. Monmouth, a nearby house (1820), on the other hand, displays a type like Nashville's Belle Meade, with heavy substantial square pillars and with little of the Natchez tradition in the façades.

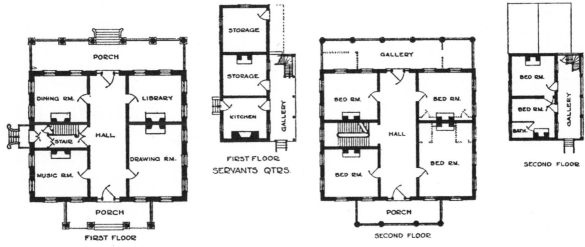

"ARLINGTON"

Adams County, Mississippi   D'EVEREUX 1840

D'Evereux can well represent Natchez in the Greek classical bracket. William St. John Elliott was interested in "up to the minute" style as was becoming his social ambition.

D'EVEREUX

D'EVEREUX is the official representative of the Greek Classical influence in Natchez. The appearance of this influence in a highly developed state at D'Evereux is unusual for that vicinity. Parts of Alabama and Mississippi show many good examples; but Natchez and the Felicianas were sparing in their patronage of it except for the Greek temple arrangement of the verandas as seen at Dunleith in Natchez and Greenwood and Elleslie in West Feliciana Parish.

Natchez, of course, was subjected to a great many architectural influences because of its cosmopolitan citizenry; but it is interesting to observe that there was little individual blending of styles. Whatever basic style a builder selected, he persisted rather stubbornly in carrying it out, regardless of how his neighbors were building. This is especially true of later homes where the masters were wealthy men who had come to Mississippi to invest their capital in cotton lands.

In D'Evereux, dating from 1840, we find an excellent monument to an imported style, alone and refreshing among other traditions. It was designed for William St. John Elliott and named for one of the owner's family lines. Like most other great houses of that day along the river, D'Evereux was the scene of magnificent entertaining, and was much written about by travelers and local historians.

The plan of the house is similar to that of Rosalie and Arlington, with less monumental stairs leading off the main hall and with the fireplaces on the outside walls. The original simple arrangement of four rooms upstairs and down with verandas across the opposite elevations has been altered to provide a series of pantries, baths, kitchens, and sun-porches for present day living.

In elevation the order is obviously Greek Doric, with which liberties have been taken. The columns are good; the entablature is all present with frieze, architrave, and cornice; but such details and ornaments as triglyphs and mutules are conspicuously absent. Colonnades of six columns each appear at the approach and opposite elevation, supporting the usual hipped roof, crowned with a cupola. The cupola shows a late and full development with promenade and wood balustrade. The entrance doors at each end of the great hall are flanked with side-lights and transoms, and are recessed with simple, substantial treatment. The second story porch is eliminated on the approach front and a small cast-iron balustraded balcony over the entrance substituted, another feature seldom seen in Natchez or further south. However, the second story porch, a real necessity not usually sacrificed to vanity in this warm country, reappears on the opposite elevation. Simple fenes-

tration augments the importance of the white pillars of D'Evereux.

D'Evereux occupies a perfect site on a hillside, a considerable distance from the highway. The natural forest makes an effective, even dramatic backdrop for the formal plot of the house and its grounds. We are told that the gardens, which were of unusual size and brilliance, were planned and executed by professional gardeners, as were so many of the famous ones of the South. Here the formal plot assumed the shape of a series of terraces on which blossomed japonicas, roses, and azaleas alternating in great beds. At the lowest terrace a lake reflected the brilliance of the flowers on one side and the deep greens of the forst on the other. The lake had its old mill and water fowl, much as in our modern schemes. The front approach was by driveways which crossed and recrossed, framing formal beds of blooming shrubs and Cherokee roses.

Almost everyone who describes the ante-bellum gardens of the South begins and ends by stating that "foreign" gardeners were brought in to design them. That gardeners were brought to Natchez and New Orleans and Nashville is a matter of record, of course. They came from the eastern states and from England and France, as well, often immigrating to America to pursue their trade. They were of the same ilk as the intinerant workers in wood and plaster who traveled over the South from place to place as their services were needed; hand-workers who could execute skilfully but who rarely had been trained in the principles of architecture or landscape gardening. They quickly adapted their methods to the new conditions of the sections to which they came, and doubtless served as teachers to local workers, especially to the slaves on large plantations.

Much credit for the beautiful surroundings of the mansions is due to these imported workers, but they were not, to my way of thinking, the only designers of the gardens of the Old South. The real landscape gardeners of Natchez and Columbus and Nashville were the plantation owners who had cleared the land for cotton or cane, and visioned a homestead in the midst of their acres. Like the brick burned in kilns on the place from native clay, like the timber and stone supplied by the plantation itself, the flowers and shrubs of the garden were a part of the landscape. Many old garden encyclopedias, calendars, and textbooks dating back as far as the seventeenth century are today evidence of the fact that there was available ample prototype material of practically every famous garden in the world at that time. From these sources of information they learned of famous French, Italian, English, and Far-Eastern designs as well as gained instruction as to practical plant culture—

Natchez Vicinity, Adams County, Mississippi

MELROSE 1845

These parlors of Melrose are typical twin parlors of the Southern house. John T. McMarran wanted to keep up with his neighbors.

flower, shrub, fruit and vegetable. The lady of the manor was forever trading bulbs and seeds with her neighbors, nurseries were installed to develop wild species, and the slaves were taught gardening skills.

Part of the dream of nearly every builder of a "great house" in the South was a spacious garden to be enjoyed by the family and their friends, and quite frankly, to be seen and admired by passers-by.

MELROSE

LYING deep in a meadow, surrounded by the wild beauty of the Natchez country is Melrose, a typical glamour house of ante-bellum Natchez. The first view upon entering a wooden gate at the end of a long country lane shaded with the usual live oaks and moss is that of a quiet pond bordered by native cypress. Beyond is a broad moor rising in successive gentle slopes of grass, the shimmering verdure of which eventually ends at a formal fence bordering a woodland.

Melrose presents façades very much in character with Arlington and Rosalie; the conventional Natchez portico of simple Doric order and pediment acting as a central main façade. The motif again varies just enough to denote individuality. The usual hipped roof flanked the portico. The Georgian influence in fenestration and exteriors are ever mindful of Natchez.

Inside to the right of the hall are the drawing rooms, which are illustrated here. These double parlors, typical of all Southerners, indicate a feeling for fashion. Here the heavy fabric known as brocatelle lends sophistication and this same pattern, which has the appearance of being embossed, covers the important pieces of furniture and is the key to the gold and green color scheme of the front parlor. The dignified marble mantel is topped with a handsome mirror, while another mirrored wall decoration, sympathetic in color and texture, is of gold leaf French design

reaching from floor to ceiling. An interesting old table which bears the saber scars of war, is inlaid with delicately hued marble forming a bird design. The jeweled eyes of the birds were picked out by Union soldiers. A mirrored atazere reflects many little keepsakes of by-gone days. Rose and gold are the predominating colors in the second parlor and here, as in the first parlor, rosewood recalls the Parisian fashions of furniture. Color on the floor is invaluable to a sense of unity and the richness of the carpet adds spaciousness and affords a satisfying setting. An artist's touch can be seen in the skillful treatment of the antique chandelier—heirloom from an old mansion in New York. This chandelier is made in tiers and has for its apex a wide-lipped vase of hand-etched crystal. Each candle socket is surrounded by a chalice and hung with glass prisms.

The southern portion of the house is lengthened by a library. Qualities of Southern hospitality did not fail to penetrate the dining room which was complemented by the old punkah suspended over the dining table. Imagination need not be vivid to see the grinning darkey propelling this fan during the serving of meals.

Cooking in a separate building is almost obsolete but not here, for Melrose still preserves this ante-bellum custom. An old smokehouse adds its part to the make-up of this typically Southern home.

St. Francisville Vicinity, W. Feliciana Parish, Louisiana

ROSEDOWN 1835

Greek classical influence has found its way into the happy lands of Louisiana. Here John James Audubon completed his *Birds of America,* and Rosedown library today boasts of a double elephant edition.

PART TWO

THE FELICIANAS

ABOUT fifty miles down the winding, twisting Mississippi from Natchez you cross into Louisiana into the parish of West Feliciana, as it is called to distinguish it from its twin to the east. Another twenty-five miles and your boat will land at St. Francisville where you climb the steep road to the little town perched on its red hills safe above high water. In 1860 it was the most important landing between Memphis and New Orleans, for this was and still is a rich section of the cotton kingdom. Even in the 1820's when John James Audubon took the steep road from the landing out to the Pirrie plantation where he was to serve as tutor to the daughter of the household, St. Francisville was a place of importance and the home of a score of wealthy families. Audubon, to be sure, was more interested in the birds he glimpsed as he paddled along Bayou Sara than in the planters and their families whose portraits he sometimes painted, or in their fine houses. Yet it was in the Felicianas that he found the security and encouragement which enabled him to complete his *Birds of America*. His hosts and the neighbors were kind, so kind that they invited him to bring his family there, too. After a fruitful summer, writes his most recent biographers, Audubon "had come to think of Feliciana as home; so he was to call it for the rest of his days." No matter where he went later, he never forgot it.

A strange lush country, these parishes of Feliciana, sometimes threaded by bayous and dark swamps, thick with great trees, poplars and beeches and magnolia, that are the haunts of brilliant birds and gliding snakes. People who have learned to live there are, like Audubon, more than likely to call it home for the rest of their days. Stark Young gets at the heart of its charm in a sentence or two in his *Feliciana*: "They say that when he comes into the parishes of Feliciana, a man, without forgetting to please others, may act to please himself most variously. We can conjecture, perhaps idly . . . what part in this verdure the sun, the great river flowing past might take." The first settlers called the region "Feliciana" and its inhabitants since have continued to look upon it as the "Happy Land."

The early history and culture of the St. Francisville vicinity is much the same as that of Natchez, except that it was thrust even more deeply down into French Louisiana. The influential settlers were predominantly British although many French and Spanish families lived nearby. The Americans who became masters of the cotton plantations quickly amassed fortunes, and as quickly built handsome homes. The three houses we shall visit in West Feliciana are typical of the grand manner.

ROSEDOWN

FELICIANA houses, for the most part, sit back from the highway several miles. The Rosedown plantation house is like this. I rode along a low white fence through the forest until there opened up in front of me an avenue over a hundred feet wide, vaulted over by the arms of giant live-oaks and draped with gray moss. Through a large wooden gate I gained access and started down the avenue. Occasional flickers of sunlight penetrated the gloom to spot the carpet of green moss and ivy, and shafts of sunlight seemed white pillars supporting the roof of the forest. I passed cross avenues leading off into the gardens that flank the road, and had glimpses of great white statues and enormous urns among the trees. It was almost like walking down the nave at St. Peter's in Rome and catching sight of beautiful chapels to right and left.

I knew Rosedown was at the end of the avenue, but I was unable to get a clear view of it through the live-oaks. No matter where I set my camera, I could get only so much—just enough to stimulate the imagination. The view in the illustration here is from the only available vantage point, and at that I imaginatively lopped off some of the shrubbery. Besides the live-oaks and pecans that form the background of the planting, there are junipers, azaleas, japonicas, bay-trees, lavendars, and all sorts of unusual tropical plants.

The gardens at Rosedown, although they are ghosts of what they were in the great days, are worth a trip half across the Continent to see. Along their faded paths and by-paths appear all sorts of pleasant surprises—niches, footbridges, trellises, and arbors—and formal statues representing the seasons and the continents. Past miniature flower gardens and orchards, one comes upon the playhouse and schoolroom where the children were once tutored by Audubon. No wonder the library at Rosedown contains an Elephant Edition of *Birds of America*. It was pleasant to have a tutor-guest turn out to be a famous man, and well worth subscribing to his book even if each number did sell for two guineas.

Rosedown was built by the Turnbull family in 1835, and is occupied today by the descendants of the first owner. There is a firm tradition that the gardens were laid out by a French landscape architect, and one can well believe that it is true. I know of no evidence regarding the designer of the house; but I can conjecture much about him. At first sight of Rosedown one wonders if the builder was not answering a challenge to apply the ancient orders of architecture on an already homogeneous mass. The central feature, two full stories in height, with the gabled roof proper housing the veranda and giving generously of its "cover," is typical Louisiana style. It is what would be expected of an evolution of the old traditions with added knowledge of classical motifs. But the one-story portions which flank it on each side create an additional interest in composition, and upset one's equilibrium in trying to account for the germ arrangement—a Tidewater Virginia manor house in a Louisiana wilderness. The whole is a very impressive mass in silhouette as it grows out of the midst of a moss-covered garden at each end and emerges into the central colonnaded motif to terminate in a colossal avenue of trees. Rosedown is a monument to the versatility of the builders of the white-pillared house of the Deep South.

The central two-story veranda is framed by plain Greek Doric columns, superimposed in the best Louisiana manner. They are solid shafts of wood, each cut from a single cypress log. The house, too, is entirely of frame construction. The entablature is heavy, but in accordance with the best textbook proportions, with triglyphs and all members present. The one-story-wings flanking each side of the veranda continue in the same order of architecture with variations of pilasters, a motif seldom used in the South. These wings, if detached, could easily pass as little temples complete in themselves.

Balustrades are very generously used for parapets and porch borders. All of these details are well applied and are indicative of a wide variety of traditions available to the designer. There are the Greek details of the orders, the Classic pilaster and balusters, and the exquisite Georgian influenced fanlights and trim. Any one of these, done as well as at Rosedown, would be considered an architectural accomplishment for the average builder.

In the collection at Rosedown, of really good old furniture and fittings, one finds that the essence of its charm lies in the diversity of its materials. The Turnbulls were widely traveled, and from every corner of Europe and the Orient they gathered these treasures for their home. Today every piece of furniture and every ornament is still in its accustomed place. Even the original Longfellow pattern wallpaper in the hexagonal hall holds its own and has been dimmed very little by the flight of the years. Bronze chandeliers fitted with lamps for burning whale oil illumine this hall. The rare portraits, painted by the great American artist, Thomas Sully, fairly sparkle in their freshness. Before the parlor fireplace, with its mantel of grained black Carrara marble, andirons and implements of hammered Colonial brass, is a cross-stitch fire screen embroidered by Martha Washington, presented to the Turnbulls by a relative of the Custis family. A whatnot with mirror back is a decided high-light in the parlor. The rich crimson brocade draperies edged with heavy gold and crimson silk braid are held in place with gilt hands holding sprays of flowers. Linen shades, hand painted in floral design, peep beneath the curtains. The Empire mirror over the mantel reflects the Italian statuary at its base and the carved rosewood furniture made by Prudence Mallard and Francois Seignouret of New Orleans.

In the music room is a grand piano especially designed for Rosedown by Jonas Chickering, earliest of American piano-forte makers. On the mantel of this room are girandoles of extreme elegance. The library contains rare contemporary books in three mahogany bookcases built in France to the specifications of the Turnbulls for this particular room. An Aubusson carpet in a flowing rich pattern completes the ensemble of the parlor, music room and library.

AFTON VILLA

THERE is a garden down in West Feliciana Parish, five miles north of St. Francisville, that has long been fitted with features of culture peculiarly adapted to it. Conceived by a French landscape architect in 1849 and belonging to Afton Villa, this property of Dr. and Mrs. Robert E. Lewis is on Highway 61.

The entrance, down a long avenue of live oaks interspersed with azaleas, surpasses most places I have ever seen for natural elegance. Abounding in informal beauty, this setting, carved by ages and softened by sunlight, affords a fine restorative quality for tired minds. As one winds among century-old live oaks dripping in soft gray moss, native magnolias, cedars, pines, and yuccas give all-year promise of their various shades of green. The absence of attempt at experiment in this highway-to-house entrance and the perfect complement of trees, flowers, water, and drive are so successful, one is loath to imagine one without the other. This fifteen or more acres of constant pleasure to the eye continue and finally a great wide space bursts into view. Here the grounds are dotted with live oaks, even more azaleas, and very old tree-like camellia japonicas—all bordered with spireas thrown in relief by dense woods. Now the house comes into view for the first time.

Remodeled and enlarged in 1849 for a Kentucky bride (the second wife of Mr. David Barrow) in the days of lavish entertaining and great homes, it stands as a marked example of ingenuity—notable for its size (forty rooms) and its Louisiana interpretation of Gothic architecture.

The original part (from the ballroom back) was built in the early days of West Feliciana Parish and is characteristic of that period. The front, as it now presents itself, is the 1849 addition and the Gothic part.

Afton Villa, not being a white-pillared house, would ordinarily be excluded from this book, but because it is a most important contribution to its type and because of its entourage, it would be impossible to complete a picture of Southern life and architecture without taking cognizance of its influence. One is continually reminded of the great impression things Gothic made on the Southerner while on the grand tour. No community from Cincinnati to New Orleans is free from its grip.

Kentucky, about the middle of the nineteenth century, displayed quite a flurry in the revival of this old style. The Bluegrass country, especially around Lexington, developed many examples among which was Ingleside built in 1852, Landown, 1850, and Ashland, 1857—all very good Tudor Gothic. No attempts were made whatsoever to use the Gothic influence in white-pillared houses as we find them in Columbus, Mississippi.

The Southerner was persistent in his demands for these delicate mouldings and lacy forms, regardless of the fact that few craftsmen were familiar with them, and materials were limited to wood and plaster. Occasionally a "circuit rider craftsman" would show up in a community direct from France or England, who possessed a good working knowledge of the art in the use of these local materials. Immediately, every planter would "book" his services; then plaster cornices, ceiling pieces, and mouldings would adorn every mansion in the neighborhood. Since the details were identical, it would be difficult to tell whose parlor you were visiting, if you were led "to location" blindfolded.

Few people in the Deep South, however, attempted any innovations beyond these precious details, not that they would have been too big an undertaking. No building was on too large a scale for a plantation lord in the lower valley country. (Especially are we convinced of this after viewing the ruins of the Versailles Plantation on the Mississippi below New Orleans.) The Gothic style just did not seem to suit the plantation as well as white pillars, and there seemed to be no need to develop it to such an extent when satisfactory labor was so difficult and practically impossible to obtain.

David Barrow evidently had other ideas about Gothic. This house was to be for a bride who I understand was from Kentucky. Anyway he secured the necessary craftsmen and enlarged his house; the new portion Gothic all the way through. The plan is in accord with the demands of the day. However, in consideration of the old house, which was included, there are necessary departures from customary details. The façades are Gothic tracery carved out of native cypress and represent endless labor, to say nothing of the skill of craftsmen and knowledge of designers. The predominating detail is in English interpretation with, however, occasionally a reverse curve bordering on late flamboyant, a French interpretation. There are insert porches, balcony porches, balconies, stem pillars, arches, mullions, muntins, balusters, railings, brackets and mouldings, steps, spandrels, panels and panel tracery—all carved out of logs—piece by piece.

On the interior, the stair hall carving is comparable with any English manor house of the period. There are carved doors, vault ribs, spindles, newels, casings and mouldings of all kinds done in Gothic accuracy. There are plaster vaults and cornices, plaster imitation marbles, and marble floors. A remarkable hall when viewed from any angle, and one which represents a great deal of work and very little unity. The stair hall is sketched for your inspection of these

Vicinity of St. Francisville, W. Feliciana Parish, Louisiana

AFTON VILLA 1820 AND 1849

David Barrow added Gothic fronts to this Louisiana house for his Kentucky bride. This staircase reflects the craftsmanship of Afton Villa, and is responsible for many fine Gothic details found in West Feliciana today.

carvings. On the south side of the house is a little ninety-year-old formal garden featuring English box originally laid out as a maze. Here the air is fragrant with Chinese honeysuckle hanging in a solid mat of yellow over an old summer house. Emerging from this bower, there are camellias, sweet olive, fuscatas, lilies and many other flowers contributing a witchery of perfume. The sunken garden in seven terraces now comes into view and, in this wide sweep below, great carpets of carefully mowed grass lend the calming influence of open space and sunshine. Here is a perfect setting for contentment.

In the distance are rose terraces, revived by Dr. Lewis, showing an endless variety of fine types where pink and red Radiance and Cochet predominate. A terraced garden of azaleas faces the rose garden, where the Pride of Afton azalea offers a shade not to be found elsewhere.

You would have to visit Afton Villa to appreciate fully its real value. It is like color and texture—hard to describe.

W. Feliciana Parish, Louisiana

GREENWOOD 1830

The importance of the Barrow Clan is summed up in Ruffin Barrow's mansion, Greenwood. Here, the scale is so large a man on horseback will hardly reach the veranda balustrade railing.

GREENWOOD

THE magnificence which came of great wealth through plantation prosperity reached its highest form in Greenwood. Owned for the last thirty-five years by Mr. and Mrs. Frank S. Percy, the house was erected in 1830 by Ruffin Barrow, member of one of the fabulously rich families that flourished in the Mississippi Valley in those days. Not only cotton, but sugar plantations contributed to the wealth of the Barrows, who owned mansions on both sides of the River.

Greenwood is a typical temple veranda house of the nineteenth century South, the colonnade reaching around the four façades and extending from the ground to the main roof cornice with no evidence of a second floor veranda ever having been present. This omission is probably the only example of its kind in this section. The other houses of this type, Ellerslie, only a few miles away, and Dunleith at Natchez, both have second floor verandas with wrought iron or cast-iron balustrades.

Eastern seaboard influence is present at Greenwood in the manner of the raised porch floor, or a sort of stylobate. Usually, in the Deep South, the column bases formed a pedestal which seemed to carry the load to the earth, but at the same time served as a support for the porch floor which was elevated about three feet above the ground level. Further south, as soon as the river leaves the red hills of the Felicianas, this use of the stylobate is dropped altogether and the first floor assumes a position a few inches above ground level, with the pillars springing individually from the earth.

The order is Doric with elongated shafts and an entablature of fair design which has triglyphs. Fenestration is simple, in good early Louisiana tradition, and the cupola is of importance as a functional element with full development architecturally.

The most important thing about Greenwood architecturally, as an example of the period and the locality, is its scale. Here we see a perfect example of the use of Classic architecture in a home suitable to the importance of the owner of vast Southern estates. An automobile driven alongside the entrance of the house will not reach the top of the veranda balustrade, and the twenty-eight pillars are each over thirty feet high.

The plan is conventional, with a monumental stair hall and the usual two rooms on each side, all repeated on the second floor. Gold leaf cornices crown the windows. Irish point lace curtains adorn the parlor windows; those in the dining room are Belgian lace. These exquisite curtains are more than a hundred years old. The furniture in the dining room is most unusual, of handcarved English oak in fruit and bird design with twelve highbacked chairs entirely crowned in the same carved pattern. An outstanding piece of silver is an enormous old Scotch venison dish. Brought over by the Scotch Fisher family from England (as was much of the furniture in Greenwood), this venison dish is nearly two hundred years old and has a hot water compartment beneath the silver platter.

In the parlor, white walls and woodwork make a perfect setting for Louis XV furniture. This rosewood furniture, with gilt copper mounts, is upholstered in the most handsome of all Aubusson tapestry: pale green background with a medallion in the center of dainty cupids, surrounded with a wreath of roses in pastel colors.

Clinton Vicinity, East Feliciana Parish, Louisiana

ASPHODEL 1833

Benjamin Kendrick, we are told, was the builder; but to every visitor to Asphodel, the Misses Sarah and Kate Smith seem to have belonged there always.

ASPHODEL

Well back from the Mississippi in the uplands is East Feliciana Parish, as interesting in its culture and landscape as West Feliciana, but with a distinctive personality. Clinton, the county seat, and Jackson, another small town a few miles to the west, were rich centers of plantation life in the years before the War, and today they are still rich in tradition and fine old homes. The classic lines of the public edifices belong to the same tradition that built the white-pillared residences.

East Feliciana Parish boasts many good examples of architecture, such as the Chase mansion, the Stone house, the Bennett house, and two old college campuses. These are all of noble proportions, but we have visited so many grand mansions in this book that I am afraid the reader will think that there were no good houses in the Old South except the large ones. On the contrary, many of the best ones are modest in scale; and more of these smaller houses have survived than those of the enormous structures like Rosedown. A beautiful example of the excellent smaller home is Asphodel, a mile or so out of Jackson. So charming is it in composition and so indigenous to the Louisiana region that it will serve admirably to represent the East Feliciana Parish.

The road to Asphodel climbs up from the creek bed through a thick virgin forest. At the top, the house emerges from the trees and shadows as if by magic—one thinks his eyes are playing tricks. The wide veranda across the entire front of the central portion has a colonnade of Doric shafts in excellent proportions that spring from high pedestals. The veranda, some four feet from the ground, seems to be suspended between these pedestals which totally ignore it and attend to the dignified duty of supporting the pillars. A typical Louisiana note is the slender wood balustrade which borders the veranda and helps to frame the simple French windows opening onto it. The larger dormer and high gabled roof over the central portion express a "hidden" second floor approached by a small interior sneak stair. A wing flanks each side of this central portion, retaining the same floor height. They are approached by separate steps from the outside, and have small stoops crowned by pediments which are in turn supported by small well-formed Doric colonnettes.

The plan displays a simple germ arrangement of Southern French Louisiana type; four rooms square, opening onto front and back verandas, with no hallway. The wings seem to have been added later, but they make for a satisfactory whole as they have small balcony connections to the center feature on the front and a common veranda on the back.

Asphodel was built in 1833 by Benjamin Kendrick, but shortly passed into the possession of the Fluker family. A quiet little home like Asphodel may well represent the parish because it so completely symbolizes the unity of the past and the present in the life of the Felicianas.

The two spinster sisters who own and occupy this home suggest, by their unusually simplified way of living, the nineteenth century antecedents. They feel there is no occasion inviting enough for them to leave this little place of refuge, hence their complete isolation from the lure of the rest of the world. In the thirty-five years of the "late Victorian period" of their lives, one has left the home only once and the other not at all. They delight in the serenity of those quieter days. The old-fashioned pot plants border the front porch, and a few minutes conversation with Miss Kate explains the reason for this nearby flower garden; her health is failing and she cannot waste her remaining energy in the yard. There is no man around to keep up the fences, and since the cattle running at large would not spare delicate flowers, these sisters enjoy their garden on the porch.

I found Miss Sarah, the other sister, too busy with her domestic duties to receive me. It seems that on this special occasion a calf had been born off in the neighboring hills about three or four weeks before and since the mother had been coming up to be milked every day without the calf, they were reasonably sure it had died. On the very morning of my visit, the calf, having gained the necessary strength to climb the hills, came back home on its own feet, giving Miss Sarah new joy in the added cares of her little world.

In the course of my conversation with Miss Kate, a big, black, red-comb hen stepped into the parlor and in her busy efforts to find a nest, my attention was called to the fact that the house was not screened (much less protected with the window guards of our modern homes); when I asked what they did about the mosquitoes, she laughed and told me they did not have mosquitoes or flies. Truly, the Lord has been kind to these little women.

By no means do I want to leave the impression that these sisters are living in ignorance, for they are charming cultured women of the old school, receiving (almost daily) intellectual people. They have choice autographed books and, in spite of reduced circumstances, there is much evidence in this home of their rich background. There is a certain gaiety and lavishness about the carved rosewood parlor furniture, the gold leaf mirrors over mantels, the marble top tables, and old secretary, and the gold cornices over windows.

This home originally featured double parlors that

were furnished exactly alike. The sliding doors still give evidence of this. One room is still a parlor. The other, however, is now used as a bedroom and is furnished in very handsome old mahogany. The bed has a half tester and the bureau a serpentine front.

The Victorian dining room can boast of twelve antique Venetian glass finger bowls, milk white plates and bowls, dishes with hens on the covers and other characteristic touches of the age.

Thanks to the "flowery fields of Asphodel" and the Misses Smith, the benefactresses.

Mobile, Alabama

GENERAL BRAXTON BRAGG HOME

When John Bragg built this house, long before the War-between-the-States, he held to the traditional "T"-shaped plan of Mobile, but we note the decided rebel trend toward the classical proportions of the orders of architecture.

CHAPTER V

MOBILE AND THE ALABAMA BLACK BELT

CHAPTER V

Mobile and the Alabama Black Belt

I LIKE to think of the southern Alabama country as a little principality nestled safely away in the Deep South, surrounded by natural wilderness on three sides and the secure waters of the Gulf of Mexico on the fourth. It is evident that the orderly processes of pioneering in the Southwest worked well along these lines—Mobile with an ideal harbor developed into a seaport, and the Alabama River system, stretching its mighty tributaries for hundreds of miles beyond and inland, furnished natural transportation and communication to one of the richest sections of the cotton kingdom. These two contributing factors, each with its own traditions and influences, joined to create our little empire which evolved itself finally into the State of Alabama. Not until the coming of the railroad did its social and commercial life share their plight with the rest of the South.

We have been following the southwest cavalcade of the American pioneers—across the Alleghenies into Tennessee and Kentucky, on past Nashville and the Tennessee Basin, southward along the Natchez Trace and finally reaching Natchez and the Feliciana country, whence they were met by the French and Spanish civilizations and their march ended. Where the Natchez Trace crossed the Tombigbee headwaters (mentioned in Chapter 3) an important branch of this cavalcade turned southward into Alabama and, joining with other important branches from Georgia and beyond, found its way via the rivers on to the Gulf. Although the Natchez Trace route met outposts of the French and Spanish civilizations some two hundred miles north of the Gulf of Mexico, along the Alabama route little of these elements were encountered until they reached Mobile itself.

Mobile seems always to have been, but until the nineteenth century it was no more than a trading post, with its few crude houses and its friendly Indians' huts. It was first mentioned in American history twenty-six years after Columbus sighted the New World. In Bienville Square, in the center of the present city, we find engraved on a granite cross, this dedication:

"To Jean Baptiste Le Moyne, Sieur de Bien-

ville, native of Montreal, Canada, naval officer of France, Governor of Louisiana and founder of its first capital, Mobile, in 1711."

For another hundred years after this, the old landmark, Mobile, was allowed to sleep, disturbed only by an occasional transfer from France to Spain, to England, or vice-versa. In 1800, Peter Hamilton noted in his *Colonial Mobile* . . .

"As a traveler walked about Spanish Mobile, he would see little of American energy. . . . The country trade was with Indians only, and by canoes; at the head of a fine bay, foreign commerce was yet small. About the streets walked stolid Spanish officers and the vivacious French inhabitants, together with negro slaves and picturesque Choctaws."

With the culmination of the Louisiana Purchase in 1803, Hamilton further notes:

". . . there began to be seen an occasional wide-awake Yankee come to make his fortune."

The Spanish, probably by force of habit, or contrariness, continued to hold authority. By 1810, the population of the Territory had reached 10,000, as the Americans began to move in up-country. In 1813, President Madison issued orders to rout out the Spanish, and Mobile became indisputedly American. By 1820, we have it from Mr. Tanner, the Philadelphia publisher, the City of Mobile had a population of 5000, while the territory served by the rivers, 127,000; and by 1830, 10,000 for Mobile and 309,000 for the territory.

Leaving Mobile and the Bayou Country for the time being, let us look in on the Southern Alabama Country which was to furnish the other necessary factor in our little wilderness empire. Some hundred miles above Mobile the Big River divides, the Tombigbee pointing northwestwardly and with its tributaries finally reaching the Natchez Trace country, which is now Mississippi—the Alabama pointing to the northeast, finally reaching the North Alabama foothills below the Tennessee River bend and its tributaries extending on into the present State of Georgia. Between and in these two rich and fertile valleys, a large part of the land has since early times been affectionately referred to by Alabama folk as The Black Belt. For the story of this American cavalcade, we can do no better than to turn to contemporary records; In a report to the Erosophic Society of the University of Alabama in 1839, A. B. Meek reviews the period of pioneering in part, thusly:

"By our efforts, gigantic and savage forests have been changed into scenes of fruitfulness and beauty. Towns and villages have sprung up with the suddenness of a magician's transformations. Rivers, which but a few years ago rolled in unfretted majesty through wide solitudes—'hearing no sound save their own dashing'—have been converted into channels of commerce, and are now to be seen, lined with floating palaces, conveying to the sea the rich productions of the soil . . . through the deep valley the long railroad is visible, passing like a thing of life, uniting distant communities together, cheapening and facilitating transportation and travel, scattering riches around its path with the prodigality of sunshine, and giving to the immense advantages of the country, their full operation upon the rest of the world."

And of its social development, he notes:

"In the case of the Southwest, let us see how its society was formed. . . . Its progress at first was slow and resisted by the primitive inhabitants. The white men consequently, who sought homes in its bosom, like the pioneers of every new country, were adventurous and daring spirits, and the manner of life which they were forced to lead, was in a great measure, lawless, self-dependent, and semi-barbarous. But, in a little while, the flow of population became more broad and rapid. Glowing accounts of the natural advantages of the region attracted emigrants from every section of the Union. The Carolinian, the Georgian, the Virginian, the inhabitants of the Middle and Western States, and the New Englander, all poured with their families into this vast and fertile field, with unprecedented rapidity. They came—in the phrase of the day—for the purpose of making fortunes and were accordingly "business men." Without much reference to each other, they settled down, wherever convenience or the hope of profit seemed to advise, and went to the laudable business of making laws and fortunes. If to this we add a considerable amount of foreign emigration, we have a correct idea of how the Southwestern States, particularly . . . Alabama, were filled with their present population. From such materials under such circumstances, what kind of a character is it rational to suppose that such a community would possess? The purposes for which they have emigrated warrant that they will generally be industrious and practical.

They have not left their homes to seek the pleasures and embellishments of life. Profit—that profit which comes from laborious exertion—is their main object. Those virtues which follow in the train of industry—like sparkles in the wake of a ship—frugality, economy, honesty—must be theirs. Hospitality—the chief of social virtues—is taught them by the necessities of their situation. The same cause teaches them self-reliance—and independence of spirit is its consequence. Intercourse, under such circumstances, must be free, unceremonious, and liberal. All being upon an equality, there can be nothing like aristocracy in society."

And of its future, he simply states:

"I have sometimes, in hours of contemplation, attempted to imagine what is to be the destiny of this vast region which we inhabit. In my fancies, I have never for a moment, separated her from the rest of the union. . . . The tide of this improvement is onward! There is no pause—no exhaustion!"

With this background let us look at the architecture: Mobile, up to its American period contributed little to a permanent architecture because it had failed to establish a permanent community life. We search in vain today for any little trace of architecture indicative of Spanish civilization, but alas, we have to be content with the Spanish moss, which at least reminds us of the nationality. The contribution of the French was more important. The Bayou country, immediately surrounding the City, reflects traces of their early life. Fort Gaines on Dauphin Island, while built by the early American, is after the French tradition. Along the downtown streets, we find a few frame and partially brick houses, reflecting the early French influence. They match pretty closely with what Peter Hamilton sums up as the popular local conception of Creole types:

"The type house known as Creole was a wooden frame filled with plaster, and with a long sloping roof, draining to the front, where a gallery faced the street. Even when the house was two story, the type was not changed. The newcoming Americans often built on the old model. The yards were filled with flowers and vegetables and a strange gesture of Latin life was the thing retained as a fence, or even a brick wall, between near and dear neighbors. This was the nearest approach the Creole made to the still more exclusive Spanish custom of the patio, or garden within the house. The Creole flower garden was behind the house; the Spanish garden—inside; while the American had a front yard, if the American happened to care for flowers."

The old Ayers House, built in 1832 on property at 57 South Hamilton Street was conveyed to Thomas

Price by the Spanish Government of 1803. The Acker Home, on Springhill Avenue, was built about 1830; and the Walsh House, also on Springhill Avenue—built in 1827. The great fires of 1827 and 1839, destroying practically the entire town, account for the sparing number of early Mobile houses.

With the ever-increasing and finally predominating American population, Mobile developed a brick town house not far different from other contemporary combinations of late Georgian and Classical architectural tradition. These houses, for the most part, were narrow of front and had their hallways to one side. They were two and three stories high and displayed many types of unusual and interesting iron balconies and porches, both cast iron and wrought. Such was the old Jordan Place on Conti Street, built in 1843, the home of Madame Octavia LeVert. The Emmanuel Home on Government Street, built in 1836, employs more Classical detail. Then there were the duplex houses which are very interesting for plan study.

For Mobile and the Black Belt, their glamorous period paralleled the middle of the century. The City by this time had developed into full commercial importance and the pride of its society was the merchant prince. The Black Belt was supplying world markets with its products and in turn receiving manufactured goods from wherever its products were used. Here the planter held forth as the leader in society. There was, however, a profound respect one for the other and an exchange of their societies resulted in an interpretation of the white-pillared mansions which compare favorably with that of other sections of the South.

The one outstanding contribution of Mobile to the white-pillared house was its development of the T-shaped, off-center hall plan. How this plan first appeared in Mobile is not known. It is not original there for an occasional example has been found in the Atlantic seaboard developments, but it was favored here to a noticeable extent. We have reproduced for study the plan of Oakleigh, the home of H. S. Denniston. The William A. Dawson Home in Springhill and the Bragg Home in Mobile on Springhill Avenue are among other local houses with plans of the same parti. In Mobile and its vicinity, as well as throughout the Black Belt and as far as the head waters of the Tombigbee at Columbus, Mississippi, we are able to detect influences and in many cases the whole plan. I have speculated at length as to the kind of living afforded by this type of arrangement. Among its main advantages is its open plan, extremely livable in a tropical climate, there being a majority of the rooms with three exposures. It affords an ideal set-up for social functions, the double parlors projecting to the front and entirely divorced from the rest of the house. This feature remids one of the early American town houses, which were built on narrow lots. It seems natural for a traditional plan to be carried over and enlarged upon in the white-pillared era. On the other hand, it had disadvantages. There is an extravagant use of hall space, because only one side "works," and also there is an awkward appearance to the off-center hall

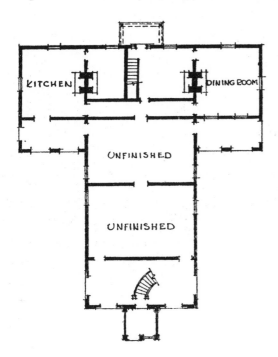

GROUND FLOOR PLAN

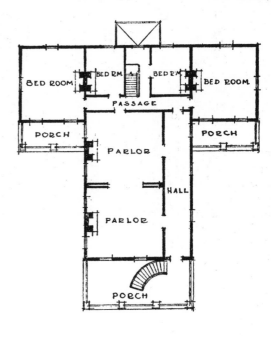

FIRST FLOOR PLAN.

OAKLEIGH

Green County (Near Forkland), Alabama

ROSEMOUNT 1832

To Williamson Allen Glover "the cupola was the thing." In Rosemount house, it is developed into many important uses. Here all other elements of composition are sympathetic. This makes for good architecture.

which is not found in the central hall plans. The builders in the Gulf Coast country, particularly in Louisiana, never did learn to use the hallway with the ease and grace of the Tennesseans.

The Bragg Home on Springhill Avenue (of which we have spoken in plan) is probably the best local example of white-pillared façades, with pillars extending the full height. Here we are reminded at first glance of the Southerner's distaste for the conventional Greek classical proportions. No particular order can be definitely identified, there are just white pillars and bracketed cornices. All in all, the Bragg house is an admirable expression of the Southerner's mode of living.

In parting from Mobile, it is well to have a last glimpse with Kate Cunnyngham, who wrote contemporarily as follows:

"Mobilians were genuine Southerners—more like Charlestonians—were friendly and 'manor born' natives—not strangers like in New Orleans."

Her stay at the first ranking hotel in America—The Battle House—was only marred by the Irish servants, who annoyed her extremely. (Good negro servants were too expensive, even for such a hostelry as The Battle House!)

Says Miss Cunnyngham,

"The merchant princes of Mobile were lords in superb villas and palatial mansions lining the noble streets,"

in favorably comparing the local society with that of Charleston and the Atlantic Seaboard. Particularly was she smitten with the "lady of the generation," Madame Le Vert, who seemed to capture the attention of every visitor to Mobile. In parting from Mobile for Montgomery, this visitor notes—

"In 46 hours, after a pleasant sail up the Alabama River, we reached the stately Capital of Alabama, Montgomery." (400 miles by river —190 miles by land.)

ROSEMOUNT

I AM glad the Black Belt mud did not make good brick and that Williamson Allen Grover had to rebuild Rosemount. It is one of the Deep South's finest Classic compositions done in wood. Truly the delight of this old house and its surroundings is due to the fact that they are both moulded out of the material at hand and without restraint as to plan type traditions.

Rosemount was built in 1832, a wedding present from father to son. Today one of its most pleasing features is the apparent ease with which it embodies what is best in the design of late Georgian and classic Architecture and furnishings. I am thinking in the first place of the germ arrangement of plan.

The reception hall, as it receives one in fitting style from the spacious veranda, is flanked by parlors right and left which, by means of enormous cased openings, make for one large room. This ensemble of hall and parlors in turn spills into a thirty-foot by sixty-foot cross hall, running the full width of plan. Across this hall and on axis with the entrance foyer is the state dining room which in turn is flanked by bed rooms on each side. Simple, unpretentious staircases lead off this cross hall to the second floor where a fair duplication of general arrangement occurs in six enormous bedrooms. The third floor cupola is appropriately dubbed "the music room." In plan arrangement Rosemount is a glamour house of the Black Belt. It shows Mr. Glover as a man of wealth, culture and social position; it reveals his efforts to

build a house to suit his particular needs, rather than one to conform with the conventional plan of the day.

Allen Glover, the Elder, born in 1770, had come with his wife Sarah Sarana Norwood, from Abbeville District, South Carolina, to the territory, Alabama, in 1818. Beauty that money cannot buy has been called gypsy riches, and this is what the son of Williamson Allen Glover discovered in the ultra-rich Black Belt of Alabama more than a century ago. To him this soil was a thing worthy of conquest and his cultural background fortified him with certain principles which were always reminding him that "to establish his domain he would have to root himself deep into it."

Rosemount rests on a star-shaped knoll, and forms a natural pattern of land planning which includes a large lake, giant magnolias which border an old-fashioned formal garden, orchards, pasturelands, farm groups and slave quarters. All this may be viewed from (and was the inspiration for) the crowning glory of Rosemount—the cupola. From this cupola not only may the entire estate be seen, but three neighboring counties and the distant shores of the Tombigbee and Black Warrior Rivers. This sight is a delight to all, and the serious, thoughtful visitor often lingers as in a trance, after the crowd has descended to the garden below. The inevitable conclusion to this guest's contemplation is that the Williamson Allen Glovers were indeed fortunate in establishing their stock in the soil of the Black Belt.

We have viewed how the pattern of life here has affected interior arrangement and land plan; to have failed to achieve this harmony in the façades of Rosemount would have been a major catastrophe. There could have been no such failure when the builder was Williamson Allen Glover. It was impossible with a man who knew honesty of purpose, and who followed it with an understanding determination of execution. He planned that Rosemount would grow out of this knoll; that there would be one-story porches, two-story portico, a simple hipped roof carrying higher and higher still to a center deck and there— the glorified cupola—theme center of the group, crowning all. Everything from every façade seems to point with pride to this prize of all cupolas in the South. Even the six white Ionic pillars of the portico seem to relinquish their title under its spell. Rose-

from James Monroe, signed 1820; from Andrew Jackson, dated 1829; and another from John Quincy Adams to Amelia Gennerick, 1825 "in the district of St. Stevens" (which was the earliest and most historic of the State's capitals). A copy of the magazine, *The Ladies' Monthly Museum Cabinet of Fashion,* is an interesting little issue in this place of safe keeping.

On an old sideboard stand carved silver candle sticks, a wedding gift from the King of Sweden to the great grandfather of Dr. Keith Legare, the present owner. A window treatment, elegant in compliance with the taste of the period, is handmade lace curtains from a convent in France. An exceptional piece of furniture is the tip-tilt wall table with top hand painted in roses on glass, backed by black velvet. On the second floor are six large bedrooms featuring interesting old mahogany beds with bureaus to match.

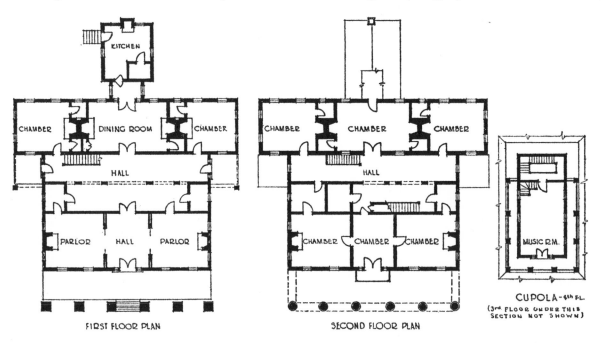

FIRST FLOOR PLAN   SECOND FLOOR PLAN   CUPOLA—4th FL.
(3rd FLOOR UNDER THIS SECTION NOT SHOWN)

ROSEMOUNT

mount folks also evidently felt its importance for there was an elevator carefully placed and seldom idle because up in the cupola was a large music room and below a house full of daughters. All together, from every façade, the silhouette is pleasing and everywhere there is good architecture; the orders, fenestration and details are in harmony with materials as full cognizance is taken of their relative values.

Inwardly, architectural detail is also in full agreement; good plaster and woodwork, marble mantels with pier mirrors, Victorian, Empire and French furnishings, collections of rich handicraft, many gay objects d'art, fine china, silver hurricane shades and the like, make for a general effect of comfort and luxury.

In one of the chimneys is a built-in iron safe in which the present owner still keeps valuable papers. Among these yellowed documents are land grants

These range from early Georgian to Greek revival. Most of the furniture was shipped from Boston by water to Mobile and up the Tombigbee River on William Glover's own boat. In each of the twenty rooms is a bell pull and each bell registers with a different tone in the servants' waiting room, as was frequently the arrangement in other houses of this period.

The drawing of the old garden is practically self-explanatory; it was landscaped, probably in the 1850's, by an Irishman named Hapt and a certain Givos must have assisted, as his name is written in the *Garden Calendar.* This *Garden Calendar* was published in 1787 in Charleston, South Carolina, by Robert Squibb, nursery and seedman.

Varied and many bulbs are footprints of the women who walked the paths of this old garden.

Roses reign in the back hedge, featuring Seven Sisters. Lady Bankshires decorate the cedar summerhouse at the entrance of the garden. Other old fashioned flowers are found in this enclosure: sweet olive, flowering quince, kiss-me-at-the-gate, honeysuckle, Confederate jasmine, magnolia, fuscatti, magnolia grandiflora and Japanese magnolia. There are japonicas, cape jessamine, crepe myrtle and perrennials inspiring gayety and arousing keen appreciation in every month of the year—fleeting but radiant! Leading off from this garden is a huge magnolia tree whose lower limbs have touched the ground and become rooted in the rich soil, producing a circle of large trees around the old one. Would that we had more of these old formal gardens—shrines to beauty under a Southern sun. To the side of the garden are olive trees, remnants of the Napoleonic Colony established in the vicinity of Demopolis by virtue of land grants from the United States Government in 1816-17.

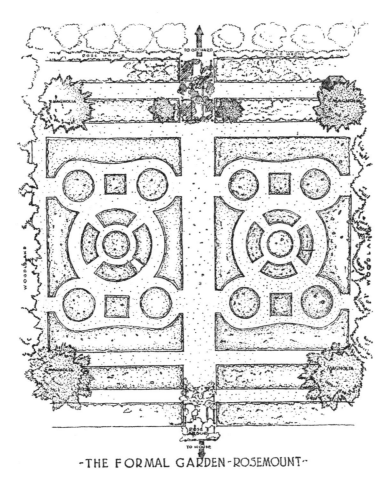

-THE FORMAL GARDEN-ROSEMOUNT-

Every Southern mansion house catered to a formal garden somewhere within the ensemble. Famous English, French, Italian, Spanish and Austrian gardens all contributed to its patterns, many of which will today be recognized by close observers.

[143]

Tuscaloosa, Tuscaloosa County, Alabama

THE GORGAS HOME 1829

Ornamental iron and white pillars were the themes architect Thomas Nichols brought from Philadelphia, Pennsylvania, for the beginnings of the University of Alabama. The Alabama folk furnished the floor plan.

A NUMBER of old structures have been selected by the Historic American Building Survey as worthy of the following certificate, one of which hangs in a frame on the wall of the Gorgas House.

The Department of the Interior

Washington, D. C.

This is to certify that the Historic building known as the Gorgas Home in the County of Tuscaloosa and the State of Alabama has been selected by the Advisory Committee of the Historic American Buildings Survey as possessing exceptional architectural interest and as being worthy of most careful preservation for the benefit of future generations and that to this end a record of its present appearance and condition has been made and deposited for permanent reference in the Library of Congress.

Hall. In 1879, General Josiah Gorgas moved in and after a short time, an addition was made to the house to be used as a hospital for students, who looked upon Mrs. Gorgas as their ministering angel.

Because men make history and architecture records it, we have presented throughout this book a close association of men and houses. The South has boasted of her men, though little homage has been paid their houses until lately. That these men have inevitably at some time recorded their way of life in houses and that these houses have preserved such for posterity through generations of great and near-great descendants is evidenced here.

Amelia Gayle Gorgas was the mother of six children, two sons and four daughters. Of the six, the most famous and beloved by all nations, is General William Crawford Gorgas, the world's greatest sanitarian. He entered the United States Army and rose

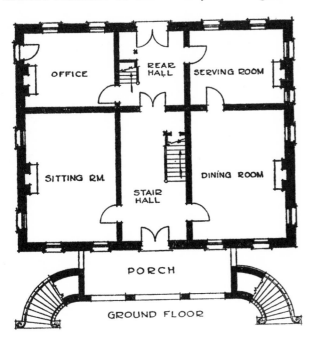

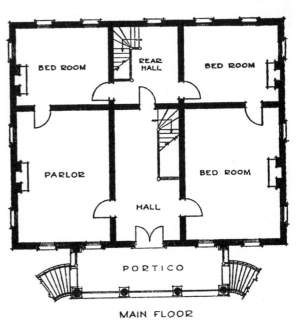

"THE GORGAS HOUSE"

This home is also listed among America's oldest college structures in a survey completed by a widely known company of New York builders. "The Gorgas Home at the University of Alabama," says the report on the survey, "is an impressive example of earlier American architecture which has retained its usefulness through more than a hundred years of National progress."

Designed by Thomas Nichols, architect from Philadelphia, in 1829 (ten years after Alabama became a State), it was first a men's dormitory and the large basement on the ground floor was used as a Mess

to the rank of Surgeon-General. Dr. Gorgas devoted his life to stamping out yellow fever. He conquered this disease in Cuba and the health of the men who built the Panama Canal was in his safekeeping; after this he brought relief to South America and to Africa. Dr. Gorgas received honorary degrees conferred by universities all over the world, and medals of highest honors from France, Italy, Belgium, England and the United States. He was knighted by King George V of England, as he lay in his last illness. At his death he was given a Royal State funeral in the historic Cathedral of St. Paul in London, with full military honors, at the command of the King and by order of the British Government. However, his body has since been brought to America where it was laid to rest in famous old Arlington. As someone has said, "he served his generation and won a lasting place in the memory of all mankind; and with all he was a modest gentleman—this physician—this soldier."

The Gorgas home was the leader for the prevailing style of architecture at the University of Alabama with its lacy wrought iron balustrade and Greek influence. Simple in plan and elevation, it presents an admirable combination of materials. It is built entirely of brick with hand-cut and ground jack arches and trim. The ground floor porch is formed of brick piers and flat arches which form pedestals for the simple Doric order of the main floor portico. Springing down from each side of this portico are two enormous circular stairs of brown sandstone with very ornamental cast iron balustrades that extend across the front of the portico, tying the stairs one to the other. The whole effect of simplicity in fenestration and floor levels, plus the cornice and the hipped roof, seems to glorify the magnificence of the stairs.

The plan has the usual stair hall flanked by two rooms on each side. Here, however, the hall is divided to permit a "sneak stair" in the rear of the house, in addition to the main stair in the front entrance hall. The general arrangement of elevation is similar to that of the Woodward House in Columbus and the Beauregard house in New Orleans, but the plan is different.

On entering this house, one's attention is called to the remarkable fitness of early American furniture. There are tables of old mahogany, an antique clock, and many little occasional chairs to lend an air of crinoline primness to the distinction established by the fine Empire sofa. Here also a built-in cabinet is filled with souvenirs of many years; a beer stein with ram's head and grape design has been much admired by connoisseurs, and is a relic of General Gorgas's father's days at West Point. At Mount Vernon red finger bowls were interspersed with white ones, and so they are at the Gorgas house. For background, the white-plastered walls add a space and light-giving touch to these rooms, shaded as they are by the branches of the famous tercentenarian, "The Gorgas Oak."

New Orleans

THE OLD SCHERTZ HOME, MIDDLE EIGHTEENTH CENTURY

Since early days, the Creole Roux family's house has been reflected in Bayou St. John. The Bayou and house have both altered their courses to keep pace with the fast changing life of the last two centuries.

CHAPTER VI

FRENCH LOUISIANA AND THE BAYOU COUNTRY

BAYOU ST. JOHN
THE OLD SCHERTZ HOME

THE TECHE COUNTRY AND BEYOND
DARBY
SHADOWS-ON-THE-TECHE
CHRETIEN POINT
MELROSE

BAYOU LAFOURCHE
BELLE ALLIANCE
MADEWOOD
WOODLAWN

CHAPTER VI

French Louisiana and the Bayou Country

LOUISIANA is different.

The sturdy American pioneers who floated down the broad Mississippi on their flatboats in the days of Andrew Jackson must have felt at home as far as Natchez in spite of the lazy warmth of the sun and the foliage more lush than they had known in Ohio or Virginia. But after they drifted past the bluffs on which St. Francisville now stands and saw the red hills vanish around a bend in the river, then they would have known themselves to be in a strange country. As they went south the expanse of water grew wider and circled in more and more winding loops; the banks levelled out and were overhung with dark trees drooping Spanish moss. Little streams and bayous that flowed into the main current led off into blue-green vistas of slow moving water. The air, even in winter, was warmer and damper than up-country folks had known.

Later on the Americans would settle along the Mississippi, the Red, the Atchafalava, and the Cane, and push along Bayous Lafourche and Teche and St. John to raise incredibly rich crops of sugar cane and cotton. They would follow the waterways that determine the pattern of life in the Lower South. But in doing so they would have to learn to live and make a living in the new region, and to adapt themselves to its demands.

It is not, however, its network of streams and semi-tropical climate only that differentiate Louisiana from most of the other parts of America. Long before the English-speaking settlers came, the French and Spanish had established themselves in the lowlands at the mouth of the great river. They put the stamp of a distinct racial and political civilization upon Louisiana, a stamp which its people have maintained with a gracious, proud indifference to the customs of the rest of America.

The key to Louisiana is its name, given in honor of the Grand Monarch, Louis XIV, who, when he became ruler of France in 1661, was the greatest sovereign in the world. With France the dominant power in Europe, it was natural to expect that New France in America would prosper in spite of the conflicting claims of the Spaniards who had pushed up from Mexico to claim the whole Gulf Coast, and of the English settlers on the Atlantic seaboard. The French staked their hopes on controlling the great inland river; therefore, their priests and explorers, both equally intrepid, planted the fleur-de-lis the whole length of the Mississippi. Permanent settlers came slowly, under the colonial policy of Paris. In the early eighteenth century John Law promoted his amazing land scheme for the French; in 1718 Bienville transferred the capital to New Orleans on the Mississippi; some of the colonists who came were assets while others were not. Nevertheless, Louisiana stubbornly grew and slowly began to prosper.

French householders settled not only in New Orleans but along the streams. They pushed up the Mississippi to Natchez, where Fort Rosalie was their stronghold against the Indians, and up the Red River to Natchitoches. Around the capital they favored the Bayou St. John country, and on the west banks the Pointe Coupée section and the country along Bayou Lafourche. A considerable body of Germans who came over under Law's guidance adopted the French way of life, and became invaluable gardeners and farmers around New Orleans. By the middle of the century the French provincials were firmly rooted in the new land in spite of all their difficulties.

They built their first homes in simple style, to utilize the materials at hand and to meet the demands of the climate. The dwelling house was always one story, raised a foot from the ground on blocks, usually constructed of split cypress logs, and its high ridged roof covered with bark or thin boards. The doors and windows were of solid cypress. Grace King, in her *Creole Families of New Orleans*, describes such a house at the corner of Ursulines and Chartres Streets, built in the time of Bienville, which lasted well into the twentieth century.

Houses of this type were found in the town in 1727 by Sister Madeleine Hachard, an Ursuline nun, who pictures them thus in her journal:

"The City itself is very handsome and regularly built. The houses are well constructed of wood, plastered, whitewashed, wainscoted, and open to the light. The roofs of the houses are covered with shingles which are cut in the shape of slates, and one must know this to believe it, for they have all the appearance and beauty of slate."

The first provincial homes at Natchez and Pointe Coupée and Natchitoches were very much on the same order. They consisted of a raised ground floor with a porch all around on which doors and windows opened; they were of wood with shingled roofs.

Simplicity did not long mark the homes of the prosperous French colonials. By 1752 we learn of an elderly chevalier in New Orleans who planned for himself a new two-story home, which was to be entered by a fine perron, with a wing ninety feet long, and with panelling, mirrors, and tapestry in the salon. Within the decade nearly a hundred fine houses were built in the town alone. Just as is the case on any frontier, the early crude homes were quickly superseded by better ones, signs of the increasing prosperity of the settlers.

In 1763 France, after a disastrous war with England, signed the Treaty of Paris, under the terms of which she ceded Louisiana to Spain and withdrew from America. The Louisianans, French to the core, found themselves ruled by Spain, ringed in by British territory, and pressed against by the restless Americans who persisted in coming down the Mississippi to the Gulf. In the end, they became neither Spanish nor British, but American, when in 1803 Napoleon, who had again got title to the colony in the chess game of European politics, sold Louisiana to the United States. A recent historian has figured that he got only four cents an acre for the actual territory in the present state of Louisiana. What Jefferson really bought was the mouth of Mississippi, an outlet for the vast energies of the great valley, the key to the history of America for a hundred years.

While European nations used Louisiana as a football in politics, life along the rivers and bayous went forward. In 1765 came the first Acadians, those sturdy exiles from Nova Scotia, whom the British had turned out of their homes. These valuable colonists moved westward from New Orleans along the bayous, especially into the Bayou Teche section. They were thrifty farmers, who turned the fertile Teche Country into a rich land. And then, on Good Friday, 1788, a taper burning on the altar in a house on Chartres Street in New Orleans started a conflagration that destroyed more than eight hundred buildings: all the stores, the most important residences, the church, the town-hall, the convent of the Capuchins, the guard house, and the prison. The city would have been in dire straits indeed had not Don Andres Almonaster, a very wealthy Spanish resident, contributed heavily to the reconstruction. He re-

placed the Spanish schoolhouse, the parish church, the hospital, the Cabildo, and the Capuchin convent. A new Spanish calaboza was erected on St. Peter Street, and other government buildings—all in the Spanish style. The transition from French to Spanish architecture in the city is summed up by Albert Phelps in his *Louisiana*:

"The old French buildings of wood were now replaced by Spanish styles of stone, brick, and stucco . . . The dwelling houses, too, were Spanish in construction. Now grew up that picturesque quarter of houses with dark, cool corridors, jealously shut courtyards, mullioned windows, massive doors, and wrought-iron balconies that make today this portion of the city unique in the United States."

It was in this French-Spanish city of New Orleans that the transfer of the territory to the United States took place in 1803. To the Americans the whole affair must have seemed very foreign and somewhat theatrical for the ceremony occupied upwards of a month. To begin with, the territory had been regained by France only some two years before and the official transfer had not been made. Therefore, on November 30th, began a round of festive ceremonies in New Orleans accomplishing the preliminary conveyance of Louisiana from Spain to France; flags were lowered and raised, toasts were drunk in champagne to both countries and to America as well. During the weeks of French "occupation" there were fêtes and balls galore. Then came General Wilkinson and Governor Claiborne of Natchez to take over the rule on behalf of the United States. On December twentieth, the second transfer took place, with more military ceremony and oratory and a dinner to which the French representative Laussat invited four hundred and fifty guests. During the celebrating that followed, there were near riots over the precedence of French and English quadrilles at the dances, while the rival factions sang "Hail Columbia" and "Enfants de la Patrie" in competition. The French regime was ended, but not the French way of life in Louisiana.

It is in relation to this pageant of changing cultures that we must consider the domestic architecture of the lower Mississippi Valley and the Bayou Country. The first Louisiana houses, simple in plan, were of two and three rooms in a row with galleries across the front and sometimes the back as well. Of one story only, they were elevated several feet on pillars and approached by wood steps. The façades were simple, with sloping roofs, gabled or hipped, and with posts supporting the gallery roofs. There were windows all around, and all the rooms opened drectly onto the galleries. The fireplaces were on the inside walls. The usual materials were cypress wood and brick, outside walls being of "stud" construction. The studs were mortised and tenoned and pegged, and the space between interwoven with wood laths, filled with brick or a plaster of mud and Spanish moss and weatherboarded over with wood. This filling was called by the early Louisianans, *briquette entre poteaux* (brick between posts), or *bouzillage*, as the case may be. The roofs were covered with hand-rived cypress shingles. The interiors were originally wood boards; later plaster walls and open beam wood ceilings were used. There were shutters on all openings.

All in all, these first houses solved the problems set by the life of the region. They were high enough above ground to keep out dampness and avoid high water. They were well insulated for the sultry heat of the long summers by mud or brick filler and the thick wood roof, while the one room deep plan and wide galleries permitted ventilation. They have been

developed three distinct types. The first was one room deep and two to four wide; the second was two rooms deep and two wide, with or without a hallway; the third was three rooms wide and one and a half deep. The first and second types had galleries on two or four sides; but the third never had galleries except on one side, all three rooms facing the gallery and opening on to it. The half rooms in this plan were directly back of the three main ones, and housed the

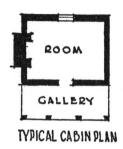

TYPICAL CABIN PLAN

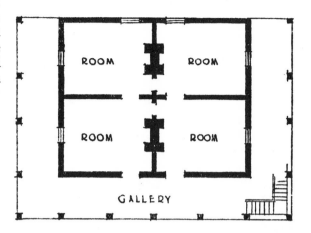

TYPICAL PLAN- TWO ROOMS DEEP

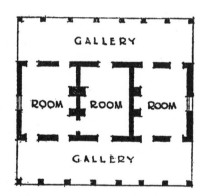

TYPICAL PLAN-ONE ROOM DEEP

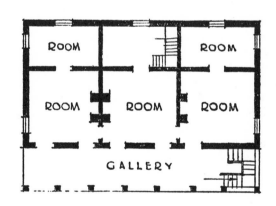

TYPICAL PLAN- ONE AND ONE-HALF ROOMS DEEP

likened to the early houses of the West Indies planters, and the likeness is real, for conditions of living in Louisiana and in the West Indies are much alike.

We have seen that the French colony prospered and grew during the eighteenth century, and that the important men built pretentious homes; yet these handsome houses in Louisiana when the Americans came into power had the same traditional elements as the very early structures. (The high pillars supporting the one story house had, in many cases, become a full clear basement or first floor and served as secondary rooms with the second floor becoming the primary living quarters.) Outside stairs were drawn under the roof and now connected the floors from one end of the gallery. The room arrangement varied from the original one room deep plan and

stairway, and the dressing and service rooms. If these houses had a second story, the same room arrangement was repeated.

Galleries became more decorative later with the classical orders coming into use both as superimposed and single columns extending through both stories; often the first floor was brick and the second wood. Staircases rising from one end of the gallery predominated. The roofs kept their original forms, but slate began to replace cypress shakes. The colors used were those of white painted wood, salmon brick, purple slate and blue green blinds. Fenestration was characterized by multi-paned sash and French windows (or doors) with Classical and Georgian detail.

In the nineteenth Century the newcomers from the

Republic who braved the lowlands of the Bayou Country bought and operated lands in the rich deltas below the Felicianas and on the west banks. At first they kept their residences well on the hills, but gradually they threw caution to the winds and moved in to the low countries, keeping rather close to the rivers, however. Thus the westward course of the Americans made inroads on the life and architecture of Louisiana. The old French and Spanish styles adopted many Classical and Georgian details; the simple and small solution took on grandeur and scale. The brilliant colors of stuccoes and slates made the old forms blossom as though they had been in bud for a century and a half; but the germ arrangement of plan and façades remained the same if the homes were truly indigenous to the land. Wherever this germ arrangement was ignored, almost certainly the life lived there was also foreign to the Bayou Country.

BAYOU ST. JOHN AND THE OLD SCHERTZ HOME

Bayou St. John has served New Orleans well in all its eras. It was an early waterway from the lower City to Lake Pontchartrain and beyond to the Gulf. Indians, before the coming of the French, discovered its possibilities and until the early years of the nineteenth century were seen approaching the city through its means. Had the English known of its source in the days of 1812 or had Lafitte agreed to show them, the story might have been different.

Bayou St. John, unlike other bayous of the Deep South, had its days of pioneering, organized country life and finally city life—a full cross-section of Louisiana civilization in both country and city ways. Nestling close to the winding level banks of St. John there was a country lane, or cow path. There was no need for roads because all commerce, ingress and egress, was by means of the bayou. Every plantation had its own craft and market barges or market boats with "brown lateen sails" manned by two or three slaves for the purpose of carrying on plantation commerce. There were also the "cushioned barges" and yachts of the plantation families, moored along the shores in the shade or sailing along in midstream—bayou families were always on the go. Perhaps one barge had a load of girls with their mammas and aunts going to shop among the treasures of Royal Street for jewelry or millinery. Perhaps a plantation household of merry children were circus bound—the boys of the party going home again to turn the great oak avenues into aerial apparatus, the groves into arenas, all the shaggy ponies into circus steeds, and to compel the long-suffering but willing household black boys to man the wagons and imitate the clowns. Perhaps M. Louis and M. Adolph LeBlanche would greet friends from their elegant craft as they came up from their estate—two hours distant on Pontchartrain—to pass the afternoon playing billiards and to meet the young girls who happened to be in town shopping. On Saturday, everyone went to town and anyone who wanted to see anyone else was likely to find everyone on the street. For the surrounding planters, town was an exchange on that day and everyone was "on the bayou"—not only the young folks in frolic, but the papas who were marketing sugar, seeing to its shipment, and laying in their stores. Those were the days—the happy days on the Old Bayou—the days of the formative period of county life which were responsible for the Old Schertz Abode and others of its neighbors lining the banks of St. John.

Moss Street now is a slick pavement where the Bayou comes to an abrupt end just around the bend. Here once, the cow path held sway. Now, with the exception of a few old timers "piddling" with boats in low wood-covered sheds, doing general commercial repair work, there is no aquatic life present, and the old Bayou's only purpose now seems to reflect pridefully its past glories and accomplishments.

The Schertz House still maintains dignity and offers shelter as in plantation days when it was the property of the Creole Roux family.

The dainty leaf ficus vine practically blankets this structure which is in a formal garden of old-fashioned magnolia fuscata, hollyhocks, dahlias, iris, roses, geraniums, plumbage, hydrangeas, sweet olive, St. Joseph lilies, myrtles, callas, "shrimp" plants and in the autumn, a riot of chrysanthemums of many hues. Two fountains also add to the setting.

The entire house has undergone some alterations (the plan of the original part illustrated here), but this has not disturbed the historic atmosphere nor its architectural significance, for the original parts are plainly marked. Here is the earliest plan type in Louisiana; one room deep and two rooms wide; front and rear gallery with steps to the second floor. This type is continued in elevations, having a hipped roof with cornices supported by heavy masonry pillars on the first floor, changing to turned wood colonnettes on the second floor. Exposed gallery ceiling and floor members here again depict the honesty of these simple builders. The type here is earlier but similar to Melrose on Cane River, and likely to be found along any bayou in Louisiana.

The first floor is a raised basement with the floor on ground level. There are the two original rooms, serving as living room and dining room while immediately to the rear are two smaller rooms with stair hall between. It is generally concluded that this part was originally the garden or back veranda, but it is also interesting to know that we shall see the same arrangement in the later houses, such as Labatut at New Roads and La Rosa plantation house on the east bank of the Mississippi. The later development into this "one-and-one-half room deep" type is interesting to watch in its evolution. Still another step in keeping with the ever-changing way of life is found in the large music room or ballroom addition to the north or garden front. Finally the simple, ever domestic, small farm house of the Bayou has become a mansion in plan scale.

The interiors are in keeping with the evolution of the plan. There are the simple, low ceilings with exposed beams, rough slate floors, plastered walls, cypress woodwork and wrought iron "H" hinges. The living room and dining room are excellently accoutred with furniture and equipment of the period, including a punkah which still adorns the ceiling. With the exception of the little "nigger" to furnish the power, this punkah still operates today as it did yesterday.

Then, in contrast with this simple eighteenth century interior, there are in the large thirty-eight foot music room pink marble floors, wrought iron balustered mezzanine, art treasures by Benvenuto Cellini and Thomas Sully, and furniture of the nineteenth century period. Altogether, such harmony prevails here, that only the most technical minded would think of exposing the secrets of the old house, and if I have disillusioned anyone, please forgive me.

Last, but not least, is the present owner; a grand person, Mrs. Christine Schertz, without whose sympathy and understanding this "farmhouse to mansion" home would probably still be just "The Old Spanish Custom House."

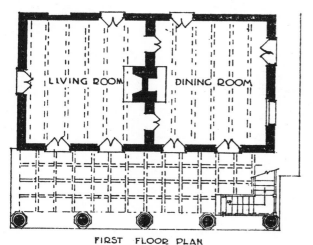

FIRST FLOOR PLAN   SECOND FLOOR PLAN

SPANISH CUSTOM HOUSE.
BAYOU ST. JOHN

Acadia Parish, Louisiana

TYPICAL ACADIAN PARISH FARM HOUSE, EIGHTEENTH CENTURY
All that is indigenous of the simple life of bayou folk is summed up here.

THE TECHE COUNTRY AND BEYOND

Bayou Teche, a circling, lazy stream, winds in a south-easterly direction for more than a hundred miles through the rich Louisiana coastal region from St. Martinsville to Morgan City and on to the Gulf. The level land on either side is patterned in plots of green, light or dark, and the red of plowed fields. There are clumps of gnarled old trees, draped with moss, and elder bushes and Cherokee roses and crepe myrtle. This is the Teche Country, and a lovely country it is in the eyes of nearly everyone who has seen it.

This section is probably more generally known as the Evangeline or Acadian Country because it was here that many of the exiles from Nova Scotia sought refuge in 1766 after their years of wandering. Their tale is known to every reader of Longfellow's "Evangeline": how the British order came to drive them from their happy homesteads, how they were separated from their loved ones, how some of them found peace again in a French land far to the south.

"On the banks of the Teche, are the towns of
　　St. Maur and St. Martin.
There the long-wandering bride shall be given
　　again to her bridegroom,
There the long-absent pastor regain his flock
　　and his sheepfold.
Beautiful is the land, with its prairies and
　　forests of fruit-trees;
Under the feet a garden of flowers and the
　　bluest of heavens
Bending above, and resting its dome on the
　　walls of the forest.
They who dwell there have named it the Eden
　　of Louisiana!"

The settling of the land and the building of homes by the Acadians was recounted by one of the exiles to her grandson, Felix Voorhies, who has recorded it.

"Shortly afterwards, we left for the Tech region where lands had been granted to us by the government. We wended our way to our destined homes, through dismal swamps, through bayous without number and across lakes until we reached Portage Sauvage, at Fausse Pointe. The next day we were at Poste des Attakapas, a small hamlet having two or three houses, one store and a small wooden church, situated on Bayou Teche which we crossed in a boat.

There the several Acadians separated to settle on the lands granted to them. You must not imagine . . . that the Teche region was, at that time, dotted all over like nowadays with thriving farms, elegant houses and handsome villages. No, it required the nerve and perseverance of your Acadian fathers to settle there. Although beautiful and picturesque, it was a wild region inhabited, mostly, by Indians and by a few white men, trappers and hunters by occupation. Its immense prairies, covered with weeds as tall as you, were the commons where herds of cattle and of deer roamed unmolested. . . . The Acadians enriched themselves in a country where one may easily become rich if he fears God, and if he is economical and orderly in his affairs, where no one may starve if he is industrious."

The old grandmother's confidence in the beneficence of the land was justified at least in her own family, for Felix Voorhies describes his father's Acadian house in the middle of the nineteenth century:

"My father's house stood on a sloping hill, in the center of a large yard, whose finely laid rows of china trees, interspersed with clusters of towering oaks, formed delightful vistas. On the declivity of the hill the orchard displayed its wealth of orange, of plum and peach trees. Farther on was the garden, teeming with vegetables of all kinds, sufficient for the needs of a whole village."

In spite of this comfort, however, the prosperous Acadians before the Civil War remained a simple people, unaffected by the grand manners and cosmopolitanism of the larger land-owners, French and American, who built fine houses. Their modest homes are still among our best examples of the true French tradition in Louisiana architecture.

Even today the Acadian region is a unique part of America, retaining much of the sturdy independence of the original settlers. Recently I made a trip deep into the Parish of Acadia, west of Lafayette, with a young attorney from that town as guide and interpreter. We traveled a whole day without hearing a word of English spoken; but my friend, as the legal counselor and political light of the countryside, was able to show me the true Acadians. During our swing around the parish we came across one old homestead occupied by an elderly man and his wife, whose story is the story of life in the Bayou Country today.

This Frenchman reared eight boys, gave each a portion of land, and built a house for each one. Lately, his major responsibilities having been discharged, the old gentleman, according to his wife, has become extravagant. Each Saturday he rides into Lafayette, gets a barber shop shave for ten cents, indulges in a "western" movie for another dime, and buys a pound of store ground coffee for thirty cents —a total of half a dollar. A short while before my visit someone had forged his name to a note for two

New Iberia Vicinity, Iberia Parish, Louisiana

DARBY, LATE EIGHTEENTH CENTURY

Francois St. Marr Darby built his country home on Bayou Teche. They also had homes in New Orleans and Paris. His wife, who was a member of the de St. Armond family, demanded a house after the French pattern.

thousand dollars, on which the local bank demanded payment. My attorney friend had explained to him that a simple suit could prove the forgery and relieve him of the unjust debt; but the old man refused. There was the note with his name on it and he would pay it even though it required a mortgage on the homestead. No man could ever say he shirked a debt. Out to the home they went, therefore, to secure the wife's signature on the mortgage papers; but the old lady refused and so saved the situation. She added, however, that if "Pa hadn't been so extravagant striking off to town every Saturday and spending half a dollar, he would have saved enough to pay off the note and not have a debt hanging over him."

A typical Acadian parish farm home is that of this old couple. It displays a number of characteristics of the early Louisiana plan described earlier in this chapter; the room and a half deep and three rooms wide type, all rooms opening out onto the gallery which houses the stairway partially inclosed at the far end. The lower story is brick with heavy brick and stucco gallery columns, while the main floor is of frame with light wood colonnettes supporting the upper gallery roof. The lower floors are paved with brick and are even with the ground. The fireplaces occupy the center wall space, leaving outside walls to light and ventilation. The roof lines here present gables at each end and include the gallery in the same slope. Oftentimes the porches of the houses are not covered in the single gable ends, but are added on with a change in roof slope, as though to express an afterthought. Such an example is Asphodel in East Feliciana.

Close examination will reveal the evolution from the early houses on high pillars, for here the brick walls have taken the place of pillars except on the galleries, and the house has become full two-story with the ground floor a useful space.

DARBY

ON all my trips to the Teche district, I had wanted a chance to study the old Darby house near New Iberia, but until very recently for one reason or another, I had to be content with photographing it and listening to the colorful legends about it. The stories chiefly concerned the present master, M. François Darby, a very old gentleman who has long been a recluse and an object of awe in the neighborhood. I give the tale as I have it from the innkeeper, but I also heard several variant versions which substantiate the main points.

François is the grandson of the builder of the house, François St. Mar Darby, an Englishman of good birth, who received the land from his father, who got it by an original grant from the Spanish. The first François married Felicité de St. Armond, a beautiful French lady of an important family. He built Darby to please her French taste. The Darbys grew wealthy and maintained houses also in New Orleans and at Paris, where they stayed most of the year. The present François Darby, third of the name, was educated in Paris, together with a brother and sister, and presented to society in New Orleans just as the War broke out. The end of the conflict left them poor and they retired, unwed, to their last possession, the home on the Teche. Unwilling to mingle with the simpler folk of the neighborhood, they withdrew to themselves and became poorer and bitterer as the years passed. They even came to quarrel with each other (so the legend insists), each living to himself in the house; for months refusing to speak one to the other, conversing by means of notes, or singing to a favorite tune such messages as "Y-o-u-r c-o-w-s a-r-e i-n m-y m-e-a-d-o-w a-n-d y-o-u w-i-l-l b-e r-e-s-p-o-n-s-i-b-l-e f-o-r a-n-y d-a-m-a-g-e." The sister was the first to die. Her favorite mode of torment was to threaten to will her part of the property to first one then the other brother—as the mood struck her. (When after her death the will was opened, it was only a blank piece of paper.) Octave, the younger brother, reduced to selling milk, further offended his brother when at long last he made friends with the townfolk, even to the point of drinking with them in the beer-parlors. He died many years ago, at sixty-five. And so François was left in the decaying house, the last of the Darbys.

A short time ago, when I was passing through New Iberia from New Orleans, I was told that old François had recently died. A filling station owner told me that the old Darby house was fast going to pieces; part of the roof and rear walls were already gone, and the storm of the night before had probably about finished it up. I hurried out to the edge of town and up the narrow, winding, moss-covered lane; and sure enough the front wall, the roof, the gallery, and two sides were all that were left standing. Through a door off its hinges I could see the blue sky. It was only a stage setting; but I started sketching to save what I could.

The entourage seemed to be in place. The oaks were as majestic as when the house was in its prime, although their moss draped low as if at half-mast. No sign of life about substantiated the rumors I had heard in New Orleans that François had "passed on."

I sketched a while then made some camera shots. The place had atmosphere; indeed, it was positively dramatic. I kept picturing the three ageing Darbys here and there as they went reluctantly about their bitter tasks. The old framework scenery swayed and squeaked in the wind as if it might well fall in a heap. I grabbed my pad to invade the interior for plans to make a restoration in case such a collapse should take place before I could get away.

The ground floor was all of brick with scant traces of brick floor under the gallery. Only a few of the brick columns remained, and those minus all their stucco; on the garden side they were completely gone with nothing but a little pile of red "bats" to mark the places. The doors and windows had lost their glass, and their once graceful blinds were half gone, half flapping in the breeze. As the plan developed on my pad, I made my way to the second floor to explore the main rooms. I was three careful steps up the flight of the gallery stairs which literally hung on one carriage, when I looked up and saw a large wooden molasses bucket on the way down. Before I could move, the head of a little graybearded man with keen black eyes and a bush of gray hair was in full view, staring me squarely in the face. From behind his big bucket he began to question me in rapid French. Then it came in English: "What's your name? What do you want here? What's that you have in your hand?"

I held my ground, probably too puzzled to move, but to my amazement, after I replied humbly to his queries, he put out his hand in greeting. "My name is François Darby, and this is my house. Make yourself at home."

He could not do too much for me. He led the way back up the stairs, insisting that I walk immediately behind him, both of us on that one little peg which held the last contact of the upper carriage to the second floor beam. Once he had gained the second floor gallery, he pulled back a shutter and bowed me into his room, the only one left with any part of a full roof over it. His father's father had built the place, he informed me, and he himself was now ninety-three.

In this room were an old, broken mahogany bed without head or foot pieces, a chest, one chair, and a long box in front of the fireplace, in which burned a slow fire. All of the other rooms were empty except for a few French books in one corner cupboard. It seemed impossible for anyone to live in such a place; but proudly he offered no word of apology, and seemingly missed none of the grandeur he had seen half a century ago. I found so much of broken beauty that I also forgot that the roof and some of the walls were missing.

Darby is simply Louisiana French Provincial style. On the main floor the center room, running the depth of the plan from gallery to gallery, served as a dining room in addition to housing a small sneak-stair. The main stair is on the gallery, the all important feature of any Louisiana plan, since all rooms on both the ground and main floors open on to it. The ground floor was merely a basement above the ground; it contained the storage space, wine rooms, laundry, and service rooms.

The illustration of Darby is a restoration of the conditions prevailing when the house was in its prime.

Although the house is small in scale, it is exceptionally well proportioned. Its openings are of an early period, long before the introduction of Georgian motifs. The simple treatments of balusters and stairs, and of exposed beams on the gallery suggest a building date in the late eighteenth century. M. Darby's own testimony tends to confirm this. The interiors are still intact enough to reveal remnants of French handblocked papers, good back-band casings, and solid paneled wood in the windows, doors, and wainscoting (1939).

SHADOWS-ON-THE-TECHE

THE Weeks homestead, Shadows-on-the-Teche, at New Iberia, is probably the best known of all Louisiana mansions. It was built in about 1830 by David Weeks, a wealthy American who established his plantation in the midst of the French Acadians. Shadows-on-the-Teche is an almost perfect example of the combination of classic traditional forms with an indigenous plan. David Weeks, his architect, or perhaps both, displayed the sort of imagination and common-sense which it is the theme of this book to admire.

Although it is only a little way off the main street of a busy town, the four-acre setting of Shadows-on-the-Teche seems several days' journey into a dense forest. The house literally springs from the blue shadows of the bayou behind it and from the dark shade of oaks, magnolias and camellias. All façades are of earthern red, hand-fettled brick and have a gray-green slate roof. Eight tall white stucco shafts, each on its own foundation, grow out of the ground and extend to the roof line across the entire street front. A light graceful entablature crowns this colonnade and in turn supports the roof, continuing across the two gable ends to the river façade, where it acts in the same capacity over a simple combination of brick and windows. The windows, of course, have blinds, which are definitely useful as well as beautiful.

The gardens, charmingly restored, are so much a part of the theme of Shadows-on-the-Teche that it is impossible to think of the place without them. Unlike most Southern gardens, which carry one slowly from formality to informality in the space of a mile or more, here the formal treatments appear in small, interesting plots at intervals, while the deep shadows of the forest encompass them. Even at noon there is shade. Every detail is an important relation to the whole. Like a painting by the talented hand of the artist and owner, it is a true example of unity, harmony and balance, perfectly adapted to its use, climate, house and surroundings. It reflects the intellectual interests and intimate home life of its owner, for it is an atmosphere for study, painting, gardening and social gatherings. It possesses qualities which all gardens should have. However, the planting is completely in harmony with the climate of the Deep South. The result—the coral house with its white columns—is serenely at home in the native bamboo, live-oaks, and banana trees.

In Shadows-on-the-Teche, we find a typical Louisi-ana French Provincial plan type, totally void of any cross-hall or wing pavilion germ of the Atlantic Seaboard influence; simple in form, rectangular, three rooms wide and one and one-half, or two rooms, deep. Wide lower and upper galleries extend the entire length of the street façade, and at one end the usual open stair, framed in jalousies, or screens of blinds, serves as the main connection between floors. The ground floor houses the minor rooms, and is only one step up from the garden. Brick and marble make excellent floors. Opening on the porch through spacious doors are the drawing room, dining room, and a room now used as a hobby shop. Just back of the drawing room, commanding the central location and overlooking the Teche, is the loggia, serving both as hall and closed porch. Kitchen and service connection with the dining room are well located on the right.

If we ascend to the second floor by means of the service stair leading off the loggia, we find ourselves in the library, a central room around which are grouped two bedrooms, a bath, and a double studio. The studio and bedroom occupy the space next to the porch and open onto it by means of large doors, exact duplicates of those on the first floor, thus making it possible to return via the porch stair to the ground floor. Fireplaces furnish ample heat for the large rooms as well as contribute to the livable quality of them. Extra thick walls of solid masonry afford deep jambs for the interior doors and allow heavy-shadowed reveals on all outside openings. Well-formed cornices are used in all the rooms and do much to keep the ceiling height under control. Early woodwork of the period in which the house was erected is still in use, and the furniture and equipment are much like those originally installed.

Here is a home which has served the needs of a Southern family for more than a century. Its present owner, Mr. Weeks Hall, is the fourth in direct line to live in it. Beautiful and serviceable when it was built, it is equally so today.

The plan could be adapted to modern living in many combinations. The workshop and service rooms of the first floor may be converted into den, conservatory, men's room, or children's playroom. The second floor can be arranged as three bedrooms and two baths around a family living room; or two groups of three room suites will work readily. Additional bedroom space is available on the attic floor.

New Iberia, Louisiana

SHADOWS-ON-THE-TECHE 1830

One would scarcely imagine the house of David Weeks to be on Main Street of a busy town of moderns. Just through a narrow gate in a tall bamboo hedge, like Alice's "Looking Glass," is another world.

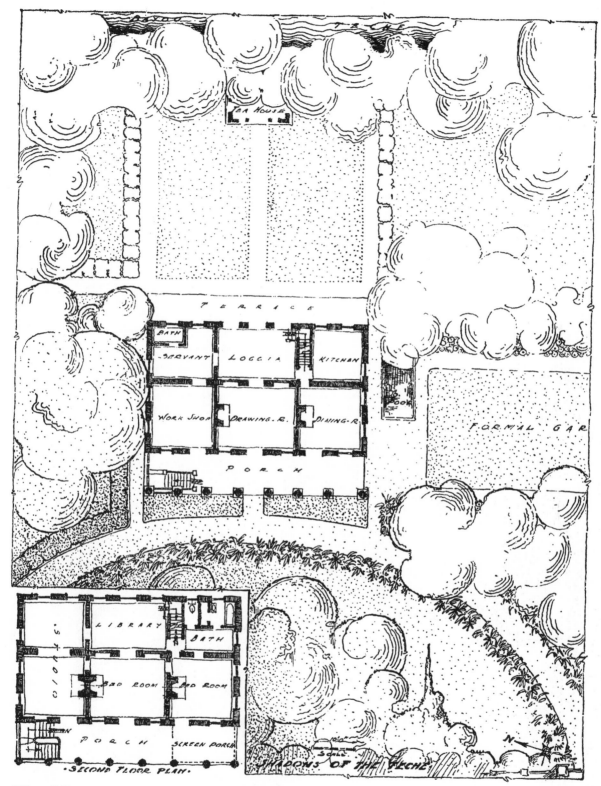

The Plan and Gardens

SHADOWS-ON-THE-TECHE

Although there was once a large plantation group here, the old house and its entourage seem to have always been just as they are today. There is fascination in exploring these several gardens. Though separated by enormous bamboo hedges, they are at the same time true examples of unity, harmony and balance.

Sunset Vicinity, St. Landry Parish, Louisiana

CHRÉTIEN POINT 1831

The home of Hypolite Chrétien is vacant now except for the folklore and memories of Lafitte, the pirate of the Gulf.

CHRÉTIEN POINT

THERE are rare times when I have found, in any corner of the South, such antiquity as is displayed at Chrétien Point. Many things happen to alter or reduce our opportunities of portraying ante-bellum life through its architecture: a house may have undergone a restoration, either sympathetic or stripped of all its character; the original furniture and equipment may be gone, or replaced in part with enough to destroy the atmosphere from within; the cultural background may have been lost and forgotten; the original families may have been supplanted by new ones who may be sympathetic only in part, or not at all. Any one of these would make the task difficult, but many of them have descended on Chrétien Point. This old homestead is vacant and lonely now (1939) . . . Like John Greenleaf Whittier's . . . "schoolhouse by the road—a ragged beggar sunning"; in a lonely field—friends all gone—out of harmony with modern trends—stripped of economic and social contacts and designs—too lazy and tired to care much—just "sunning." Yet for all its desolation, I would name Chrétien Point Plantation among my favorite houses of the Deep South. Perhaps this loneliness and stillness causes one to think of legends; to become more determined to dig out such secrets.

A few miles from Sunset, St. Landry Parish, Louisiana, and ten miles from the old town of Opelousas, is Chrétien Plantation. Hypolite Chrétien, the great-great grandfather of the present owner, Mrs. C. A. Gardiner, wanted a well-made, "good" house and this plantation home stands today—just that.

There is a story told of an incident occurring not long before the beginning of the decline of the Chrétien plantation. Hypolite Chrétien, old and nearly paralyzed, had just received the news that a band of Federal soldiers under General Nathaniel Banks had met and defeated a small detachment of Confederates not a mile from his home. Knowing General Banks' reputation as a plunderer, Chrétien felt sure that his home would be burned to the ground. When the officers dashed through the gate, they saw the stooped figure of Hypolite move slowly to the balustrade and, as a last resort, make known by a sign that he was a Mason. General Banks returned the salute, and the house was saved.

I like to think of Chrétien Point as a sort of "Rosetta Stone," for records found in this old house have been the means of translating many of the early building codes. One document is reproduced in Chapter 8, for the purpose of acquainting the reader with early methods of designing and construction. Yes—Hypolite Chrétien did things well in his day!

Chrétien was a friend of Jean Lafitte, the pirate of the Gulf, and this friendship is responsible today for many wierd tales of buried treasure at Chrétien Point Plantation, which Lafitte is supposed to have used as an inland headquarters. Here I am faced with the fact that few who have written of Louisiana have neglected Lafitte. Although he was hardly the builder of white-pillared houses, he, like "Big Harpe" and "Little Harpe," and Tom Mason, pirates of Natchez Trace fame, was a colorful character, adding to the glamour of the deep south. So I, too, feel the urge to dwell on this patriot and pirate whose whole life was *un grand peut-être*. From *The Mystery of Lafitte's Treasures*, by Frank Dobie, I gathered that some historians have glorified him as a patriot; others have denounced him as a pirate and cut-throat. He, himself, always claimed he was a gentleman smuggler and privateer. The romance of this outstanding character lures and fascinates one to read on and on, hoping to arrive at something definite regarding his life. Yet he remains a mystery, for very little which can be proven true has come to light. His birth is a mystery; his life on land and sea is a mystery, and even his death is a mystery. For more than a hundred years, Lafitte and his treasures have made fascinating conversation and reading. In New Orleans and the Deep South, his name was probably the most familiar of any man of his age, and as long as even one of his treasures is undiscovered, the associations will remain a potent monument to keep the name of Lafitte green.

The most tragic story I have gathered, however, concerning this plantation and its five hundred slaves, is the loss of the fine collection of antiques. These were moved one day to a hotel in the little town of Jannings, Louisiana, the last Chrétien having ventured into the hotel business because debt forced him to quit farming. Soon after, the hotel was destroyed by fire and with it the plantation house furnishings. So Chrétien Point was stripped of its priceless furniture and portraits, leather-bound books in French and English, and gorgeous glassware and silver. There was in this home much more than the usual possessions of these old mansions, for the Chrétiens were noble people and it is not legend that many gay parties were held within these walls. Walls that stand here today as if in wordless contemplation and mute grieving for the treasures of a century past.

While similar in plan to Shadows-on-the-Teche, Chrétien Point more than any other house I have found in Louisiana was held firmly to its true Louisiana tradition, at the same time taking on a full measure of the invading architectural styles.

Let us consider the various elements represented

Natchitoches Parish, Louisiana

MELROSE ON CANE RIVER 1833

True Louisiana types often reflected the candle-snuffer towers of Normandy. Here is the most lovable and restful spot in Louisiana.

here that are typical of early Louisiana architecture: First, there is the plan, the three rooms across, one and one-half deep type; next the pillars, which spring from the ground and support a shingled, hip roof framed to include veranda and house in one characteristic sweep; then the simple gallery with a wood balustrade, and the exterior stairways, which have long since disappeared. Next, the chimneys are on interior walls so as to preserve the previous outside space for ventilation; and last, the cool, brick floors which were even with the ground.

The plan having been carried out in true functional order, the elevations then took on the refinement of the Georgian influence for its fenestration, and Greek classical influence for its gallery. Placed in eighteen-inch reveals of massive brick walls, in the deep shade of the veranda, these circular-topped windows and doors absorb so many shadows that their reflection is coal black. Framed by deep colored blinds and red brick walls, the house was a blending of harmony envied and copied over the country-side.

Today, the house, though neglected and forgotten, shows little sign of decay. The great white pillars and woodwork are absorbing the dingy grey of the Spanish moss, the green blinds have taken on a slatish tinge of blue, and the soft earthen red brick walls seem a blend of the two, while the aged cypress roof and gutters assume a neutral color as though from profound respect. Here harmony of color goes even one step farther than the architectural element to make of Hypolite Chrétien's house a thing of rare color and worthy of admiration.

Three large panelled doorways lead from the first and second floor verandas to their respective rooms, each carefully panelled and beaded and in excellent taste. Within, the casings and ceilings are of cypress, while the verde antique marble mantels, with black onyx caps and shelves are unique.

Those who pride themselves on being modern should read the specifications (given in Chapter 8) in Hypolite Chrétien's contract for venetian blinds! Original plans and specifications of Chrétien Point will be found there also. The reader will be interested to compare 1831 with the present (1939).

MELROSE PLANTATION

Passing the head waters of the Teche, on northward to Red River Valley, thence westward past Alexandria, one comes to Cane River, and Melrose Plantation.

Cane River, at one time the main channel of Red River, but now broken off from that stream, flows through this neighborhood which is the northernmost boundary of the early French Colonial civilization in Louisiana. Melrose Plantation is a typical house of this district, where the glamour and extravagance of the later period affected Louisiana less; for here the people remained true to the French manner of living.

Mrs. Cammie Henry who lives at Melrose has, in the many years of association with this place, caught the spirit of the French provincial life and by continued research into the past, and a determination to hold this plantation intact, helped its achitectural influence to linger picturesquely on the turbulent Cane River.

After all, it is not national borders which separate one architectural style from another, but climate, indigenous building materials. and prosperity. If the history of civilization is to be written by architecture, there must be reflected in that architecture a true solution to climatical problems expressed in the materials available and in keeping with the economic condition of the civilization. Melrose is an example of just this. It begins with the simple basic plan of Louisiana bayou country, which makes an effort to obtain a maximum amount of shade and cool air. It is one room deep and several wide, with a front veranda running the full length displaying the usual staircase. Added to this on the garden front is a service ell, with a gabled porch. A simple hipped roof covers the whole, seemingly in one grand sweep. Materials are all local—brick and wood.

As to the economic and social position of its builders, Melrose presents them as modest folk requiring unpretentious façades; a home small in scale and low in ceiling heights. Their habits and hobbies were simple and interesting. They placed the genuine comforts of life before pretense or any kind of economic outward appearance, true or false. Above all, they had an intimate knowledge of tradition which was enriched by natural beauty and native imagination.

On examination of the façades, we find the native treatment of verandas with square brick pillars springing from the ground on the first floor, and turned wood columnette above, tied together with square wood balustrades and structural timbers that are left open at the ceilings. Then there are the octagonal tower wings—traditions of the provincial architecture of Normandy. Northern France has all types of towers: round, square and octagonal; half engaged, quarter engaged and free standing; one story towers, two story and up—always beyond the chimney tops with gabled roofs, hipped roof or flat roofs. Such were the "candle-snuffer" towers of these native builders of Normandy; ancestors of French pioneers of Louisiana. Whether they were original or after-thoughts at Melrose is unimportant; to me, they are a happy part of the whole. In plan they reach out for additional usable space in a way not to cut off too much ventilation. In elevation they frame in the ends and recess the veranda which seems to make the shadow more pronounced and deeper. In Louisiana, as in Normandy, we find these towers scattered country wide. They are used in many cases just as they are here—both square and octagonal, high and low. They are used as *garçonnière* and *pigeonnier*, freestanding and engaged into verandas. The roof dormers are typical of the bayou country and most Louisiana hipped roofs.

The outside French doors to each room are cheerful and inviting and there is a sturdy livable air to the brick floor downstairs. The ceiling is low, which gives an air of cozy informality and acts as a friendly foil to the separate dependencies in the back.

On the interior there is a space-giving atmosphere to the white plastered walls and woodwork, where books on top of books line the walls of the combination living room and library.

The homelike aspect of these rooms is encouraged by lamps and tables placed conveniently near comfortable chairs and sofas, where the literary seek food for thought, for Mrs. Cammie Garret Henry is nationally known for her collection of data on early and late Louisiana history. There is feast for the eyes too, in this farm house atmosphere, for there are inlaid French consoles, squatty old secretaries, brass bowls with copper trays and tops, and a very handsome samovar. All of this ties the library with the dining room. Here one notices the inside doors, wide and heavy, featuring twelve panels all in scale with the low ceiling. In the spacious dining room with fireplace at each end, there are nine foot mantels.

One sees here how snugly the French farm house can be adapted to meet the requirements of the American plantation. This room and the adjoining library seem to be the radiating point, for there is access to the kitchen and the brick court in front and back where stairways lead to the second floor. Taste and comfort prevail here and one is wont to linger for gumbo, beaten biscuit, spoon bread, fried chicken and old-fashioned custard ice cream.

No room in the house is more impregnated with the atmosphere of the plantation than the bed chamber of the mistress. So many things in its equipment recall the busy times of farm life, as the desk bearing every evidence of daily use and a grand lot of scrap books. A comfortable old sofa stands at the foot of the simple four-post bed; the tester of this bed is lined with yellow and the hand-woven spread edged with fringe bears evidence of the industrious needlework of Mrs. Henry. The homespun character of the material selected for the draperies and slip covers fits naturally and unobtrusively into these old low-ceilinged rooms. An original Franklin stove fits in the fireplace and here, as in the other bedrooms, are appropriate old bureaus and armoires.

Throughout the house there is a subtle entente between the furniture of Early America and the provincial furniture of France. They are in accord and, for an old plantation house, no better combination can be chosen.

My first and lasting impression of Melrose Plantation House brought to mind what Frank J. Foster has said of the provincial architecture of northern France. Printed in *The Tuileriers Brochure*, May 1931:

"We must be guided by the spirit rather than by the letter in studying this type of architecture. Merely to copy, line for line, will gain us nothing. We must absorb the tradition and background of these old builders, must understand their lives as well as the houses they built. We must remember that while their culture was not broad, it was thorough; that they were intelligent craftsmen with a mighty pride in the work of their hands; that their work is both naïve and subtle; that their lives were natural but often far from dull. Perhaps if we can absorb the significance of their lives we can hope to gain the essence of their achitecture."

Behind Melrose there is a design for living. The old house has mellowed and its soft, uniform, white aspect, blended with green, is at peace with this cotton country. It is sympathetic to the banks of Cane River.

Vicinity of Donaldsonville, Assumption Parish, Louisiana

BELLE ALLIANCE 1850

The Kock seat of citizenship was established here on Bayou Lafourche although the family preferred the Seine in Paris. The house is a glorified plantation house of wrought iron and white pillars.

BAYOU LAFOURCHE

Bayou Lafourche breaks off the Mississippi at Donaldsonville, eighty miles above New Orleans, and flows lazily through the lowlands to the Gulf, a short cut through a fertile country which it serves for drainage and transportation. A poetical traveler of the 1850's described this country with enthusiasm:

"In this part of the world, where the green land is as level as the blue sea, these intersecting bayous form a network of internal navigation, as if the whole land were cut up into winding canals. This feature of the country makes it very beautiful, as oaks and elms and laurels fringe their banks; and in their graceful curves they embrace, now on one side and now on the other, crescent-shaped meadows waving with sugar cane and dotted with majestic groves like islands of foliage. . . . The boat passes villas innumerable, whose gardens touch the water, and old French villages half hid in shade, while in the distance, for every half league, tower the turreted sugar houses like so many castles."

The Lafourche region is the heart of the Louisiana "Sugar Bowl" in which large fortunes were amassed in the first half of the last century. The great planters built handsomely and lived in luxury. They had many slaves, private boats, and every opportunity for culture and social grace. The same poetic traveler from whom we have already quoted has left us this picture of one of the homes along the bayou:

"At length I approached the house. Vases of large size, containing rare West Indian plants, stood on each side of the spacious steps, filling the air with delicious odors. Crossing the noble piazza, which was broad enough for a company of soldiers fourteen abreast to march round upon it, I as the chief guest was ushered . . . into a wide and high hall adorned with exquisite statuary, noble pictures. The drawing room opened into it. This was furnished with light and elegant furniture, chiefly of Indian cane and rosewood. Everything had that undefinable air of taste and comfort, without garish show, which a poetic mind loves to dwell in . . . The library opened from the drawing room, and when I say its walls were wholly concealed by carved oaken cases, filled from floor to ceiling with all the wealth of a real scholar's book treasures in all tongues, you will understand how elegant and tempting a place it is. My sleeping apartment opened from this pleasant library, and also looked out upon the lawn."

After making due allowance for this writer's romanticism, the fact remains that the Bayou Lafourche country is extraordinarily beautfiul, and even yet enough of the "grand houses" of ante-bellum days remain to substantiate the description just given. Three admirable examples are Belle Alliance, Madewood, and Woodlawn, which we shall visit in this chapter. These are not houses essentially in the native tradition but rather those of fortunate planters whose fathers had come to the heart of Louisiana from the Felicianas or Tennessee, or farther East. These are manor houses.

BELLE ALLIANCE

Belle Alliance, between Donaldsonville and Napoleonville, was built about 1850 as the home of the Kocks, one of Louisiana's wealthy families. It was mostly used as a seat of citizenship because the Kocks remained abroad much of the time. It was very well kept, however, and even today has an air of importance in its scant entourage of a few groups of colossal oaks left from the former approaches and gardens. Its grounds formerly varied from strict formality to the wildness of the bordering forest, which included many varieties of tropical botanical treasures.

The house presents itself massively in scale. The six great square pillars grow directly from the ground and continue to the entablature, pausing to carry the first or main floor veranda at a line nine feet from the ground level. Large cast-iron bolts connect the veranda skirt line to each column and are quite frankly ornamented to emphasize their presence. The main gallery steps as they appear today are projected in front and outside of the gallery. I prefer to think they were originally placed according to the true Louisiana style underneath where they were protected and were less conspicuous. Many times where galleries extended the perimeter of the house, there would be two or more flights. Lovely cast-iron railings border the veranda, spanning from pillar to pillar across the front and returning towards the rear at each end, where they are met by a complete ensemble of iron—railing, colonnettes, and canopy—forming another shelter which continues over the veranda along

Vicinity of Napoleonville, Assumption Parish, Louisiana

MADEWOOD 1845

Thomas Pugh built his house at the same time as the Woodlawn Pughs, and liked his white pillars accompanied by a pediment.

each side elevation to the rear. The whole forms an unusual and distinctive study in Louisiana types. The house and pillars are brick covered with stucco. The simple entablature is moulded of the same materials with only a dentil course to give it a touch of the ornate. The cream and tan stucco of the deeply recessed walls, the white pillars with dark green blinds, and the black iron-work form a pleasing color scheme. I have never seen a light cream stucco so interestingly used as at Belle Alliance. These old Louisiana houses, set in shadows continuously changing in the bright sun that flickers through dense leaves and gray moss, can actually deceive one as to their real colors.

MADEWOOD

Madewood, near Napoleonville, once the home of Thomas Pugh, was built about 1845. The Pughs were for generations a large and influential clan in whose homes the social life was rich and varied. Madewood is, as a matter of fact, more in keeping with the architectural traditions of Tidewater Virginia than of Bayou Lafourche; but it stands today a practical example of the adaptation of the wing pavilion plan in the needs of the deep South. Thus modified to conditions, this type of house becomes very widely used.

In silhouette Madewood is a pleasing pile with unusually good proportions. The veranda across the two-story central motif is formed by six white Ionic shafts which support a simple entablature and pediment, thus giving this portion a decided height. This feature appears elsewhere in Louisiana, notably at Linwood on the east bank of the Mississippi and at Oaklawn on the Teche, but without noticeable success. Here at Madewood, the columns instead of following the Louisiana tradition of reposing each on its pedestal, rest on the stylobate of the Greek temple tradition. The second floor line, carried just inside the pillars, is neatly bordered by a very delicate wooden balustrade. The wing pavilions, flanking each side of the two-story portion, are found repeating the pediment, and strange to say are not monotonous in their repetition. Their simple pilaster wall treatment and light stucco motifs, connecting to the center parent motif with flat-roofed passages, carrying the same cornice lines, are commendable.

The color scheme is given contrast by the use of dark green blinds on a background of white building; thus the white seems whiter and the green greener. This contrast is doubtless accentuated because Madewood lacks the entourage of trees and shrubbery usual in the Bayou Country. I think this somewhat glaring effect would be far more pleasing amid dense shadows of live-oaks.

The interior is very fine in detail. The plan is similar to that of Woodlawn (illustrated herein) but slightly more rich in treatment with its heavy columns and cornices. The left wing contains a spacious ballroom as well as the service rooms.

Vicinity of Napoleonville, Assumption Parish, Louisiana

WOODLAWN 1845

Bayou Lafourche was too accessible to the "grand parade" to retain its French influence. Consequently, it gave way to the money lords of up-country. Woodlawn reflects this change.

WOODLAWN

WOODLAWN, near Napoleonville, is another of the homes of the Pughs, built about the same time as Madewood. Their remarkable resemblance in plan and composition make them all the more interesting because they continually tell us that these two families lived the same life with common habits, hobbies, and traditions. Woodlawn, however, by virtue of its omission of the pediment has changed its proportions to a lower mass that tends to fit into the contour of the landscape and so to "belong" more completely than does Madewood.

The veranda at Woodlawn is formed by four round Ionic columns between two square Doric shafts which all together support a wooden entablature of massive proportions. The whole is crowned by a parapet breaking over each pillar to a higher point in the center. The porch floor balcony is carried just back of the column line; the main two-story portion is frame; the end pavilions are of stuccoed brick. The whole is covered by a wood roof. These end pavilions are connected with gabled roofs which continue the roof lines of the two-story central feature. The fenestration is treated with multiple divided lights; all openings carry shutters. The openings on to the verandas, both on the first and second floors, extend all the way to the floor. The central entrances have elaborately decorated leaded glass side and transomlights. The central second story wood portion and order are white, while the stucco end pavilions are soft apricot pink, with pale, greenish-blue blinds—all

weathered together so that there is little contrast.

The plan of Woodlawn shown here gives a good idea of the extravagant living expected in such a house. The central hall is not massive or monumental but more on the Louisiana idea of the hall as merely a housing for a stairway. The parlors, library and dining room are beautifully proportioned with nice detail. The most interesting feature of all is the arrangement of the rooms in each end pavilion around an open court which supplies light and air, making all the halls little verandas rambling around from room to room. The court side in each case is bordered with continuous blinds to shut out the sun. Various dressing rooms and small bedrooms for the personal servants of ante-bellum days complete the picture.

All in all, the homes of the Bayou Lafourche region contained at the middle of the nineteenth century little of the original Louisiana life. By this time the Americans had penetrated into the level sugar lands to make fortunes and to spend them. Moreover, this country was close to New Orleans and, for that matter, to Europe. Cosmopolitanism and the grand manner stamp these remaining houses of Lafourche. It is to be regretted that the earlier houses have long since passed from the picture. The older tradition of the French colonials we have found inland, into the Teche country and beyond, where the simpler folk were less affected by the glamour and wealth of the *grand parade* of power and prestige of the Old South.

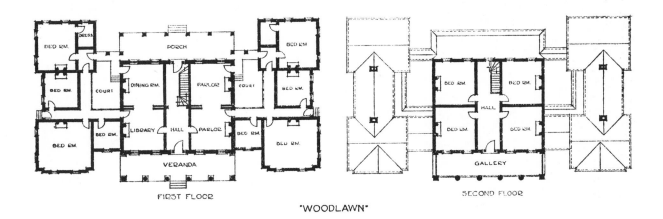

FIRST FLOOR

"WOODLAWN"

SECOND FLOOR

Vicinity of Donaldsonville, St. John The Baptist Parish, Louisiana

EVERGREEN 1800 - 1850

Here, reflected in their architecture, are all the moods and temperaments of the pioneering early
Americans who found their way to the banks of the Mississippi.

CHAPTER VII

The Grand Parade

The Grand Parade

THE Mississippi River dominated the life of the nation from the Louisiana Purchase in 1803 to the fall of Vicksburg in 1863. As James Truslow Adams has so admirably put it in his *Epic of America*:

"The heart of the new Americanism began to find its home in the heart of the continent, in the new empire of the Mississippi Valley. . . . For better and worse both, the new America was the child of "Ol' Man River," nurtured in the vast domain which had been his through all the ages. It was on frontier after frontier of his vast domain that the American dream could be prolonged until it became part of the very structure of the American mind."

From the earliest times life flowed down the stream toward the Gulf. The Indians in their lithe canoes gathered in council at the great mound where the Ohio joins the Mississippi, coming down the muddy Missouri and from the far-flung headwaters of the Miami and the Tennessee and the Cumberland. In 1545 De Soto, the first white man to see the river, crossed it at the mouth of the Red; but one hundred and twenty-eight years passed before that June day when Marquette and Joliet heard their Indian guides shout, "The Mississippi. The Mississippi!" — and turned their canoes into the mighty stream that they claimed for France and the Christian God. The next century and a quarter saw a stream of traffic moving southward with the current; swarthy trappers with loads of furs; swift keelboats manned by Mike Fink and sundry hard-drinking fellows from up the O-hi-o; broadhorns conveying settlers with their children and chattels to new homes in the great valley; flatboats low in the water full of Kentucky corn bound for New Orleans. Along the river there was always movement. On its broad waters was the outlet to the world beyond.

However, nearly all of this movement was downstream. Only the canoes and keels could fight their painful way back up against the current. Flatboatmen sold their craft and walked home up the Trace; travelers went down to the Gulf and sailed around, or else rode overland to the east. It was only with the coming of the steamboat that the Mississippi completed its vital circuit of two-way traffic. In 1811 Nicholas Roosevelt of New York built a thirty thousand dollar steamboat in Pittsburgh, launched it, and took it down the Ohio and the Mississippi in the face of a flood and an earthquake; that was magnificent. But when he brought it back upstream in the teeth of the current, he turned a page in American history. For a decade it was hard to believe that the new-fangled craft had superseded all other boats the river had known for a thousand years; then the idea caught on. By 1830, some three hundred and fifty steamboats plied the Mississippi and its tributaries; during the next thirty years the number increased by leaps and bounds. And no wonder. The cost of construction was a hundred dollars a ton, and a boat usually paid itself out in a year if no disaster befell it, no boiler explosion or wreck on a hidden sandbar. Steamboating was a desperate gamble but with such high stakes for the winners that plenty of men and money were available for the game.

The reason for the enormous profits in transportation up and down the Mississippi between 1830 and 1860 is apparent in the light of our discussion of plantation life in Mississippi and Louisiana in the earlier chapters of this book. The opening up to Americans of the virgin lands of the lower valley for cotton and sugar culture under slave labor occurred at almost the same time as the appearance of the steamboat on the river; the plantation system and the steamboat together created an extraordinary amount of wealth in an incredibly short time. The raising of cotton or sugar, but especially of the latter, called for considerable capital at the start, however, and attracted only men with money or credit. Therefore the majority of the settlers were well-to-do planters from the Upper South who were seeking an outlet for their capital and new lands for their sons and sons-in-law, although others were from more distant sections. Many of these men came without their families and lived in crude huts for a few years; some even returned with their winnings to their homes. But the majority of the newcomers liked the climate and the plantation feudalism of their river estates and so established themselves there in handsome dwellings which they occupied for at least the winter months. Steamboats were loaded at their private landings with staple crops that meant a large balance in a New Orleans or London bank. The same boats brought to their river mansions the luxuries of the great world and likewise many distinguished visitors from that world.

The wealth acquired on the plantations of the lower Mississippi became legendary. One contemporary account reads like a modern success story.

"A young gentleman whom I know near Natchez received at twenty-one years of age thirty slaves from his father and fourteen hundred acres of wild forest land on the Mississippi. He took his hands there and commenced clearing. Thirty axes do vast execution in a wood. As he cleared, he piled up the cloven timber into firewood length and sold it to passing steamers at $2.50 a cord. The first year he took $12,500.00 in cash for wood alone. The second year he raised eighty bales of cotton, which he sold at $50.00 a bale, and he also sold wood to the amount of $14,000.00 more. The third year, he sold one hundred and fifty bales of cotton and cleared by wood $10,000.00, which with $8,000.00 his cotton sold for, brought him an income of $18,000.00. Out of this, the expense for feeding and clothing his thirty slaves per annum was less than $1,800.00. The young man, not yet twenty-nine, is now a rich planter with a hundred slaves, and is making five hundred bales of cotton at a crop."

A somewhat soberer account is found in the annual report of the Agricultural Society of Baton Rouge for 1830, regarding the capital required to run a sugar plantation.

The capital invested in a plantation capable of producing, by the best management, 400,000 pounds of sugar and 10,000 gallons of molasses, worth on the plantation $23,000.00, must consist as follows:

1,500 acres of land at $50. an acre	$75,000.00
90 hands at $600. each	54,000.00
40 pair of working oxen at $50.	2,000.00
40 horses at $100.	4,000.00
Horizontal sugar mill	4,000.00
2 sets of boilers at $1500. each	3,000.00
Buildings of all descriptions	25,000.00
12 carts	1,200.00
30 plows	300.00
All other utensils such as timber wheels, hoes, spades, axes, scythes, etc.	1,500.00
	$170,000.00

The Annual expenses on the above plantation are $10,700.00 in the following items:

Provisions of all kinds	$3,500.00
Clothing of all sorts	1,500.00
Medical attendance and medicine	500.00
Annual losses in negroes	1,500.00
Taxes	500.00
Horses and oxen	1,200.00
Repairs of buildings	700.00
Plows, carts, etc.	300.00
Overseer	1,000.00
	$10,700.00

Two crops of cane are generally made in succession on the same land, one of plant cane, and the other of the second year's growth; it then lies fallow two years or is planted in corn and beans.

Gross proceeds	$23,000.00
Expenses	10,700.00
Net proceeds	12,300.00

The net proceeds, therefore are about seven percent on the capital invested.

Doubtless the profits varied widely in proportion to the skill, good fortune, and market prices, but many millionaires and a great many prosperous plantation owners appeared along the river between 1820 and 1860. In accord with the temper and ideals of the American frontier, they immediately built fine homes and maintained hospitable establishments. Along the Mississippi ran the "river road" from plantation to plantation and from town to town. Between Baton Rouge and New Orleans the road was heavily traveled by carriages and horsemen as well as humbler footmen, making it a sort of unbroken street to the Crescent City. But the favorite means of travel was the steamboat. Indeed, the great "floating palaces," as the proud owners called them, were the symbol of the place and time. We may well call the forty years before the War, so far as the Lower South is concerned, the Steamboat Age.

The laureate of the Steamboat Age, of course, is that greatest of the Mississippi River pilots, Mark Twain. He loved the great river and the white boats; and although he is humorous in this description from his *Life on the Mississippi*, he makes us feel the thrill and beauty of the scene.

"Presently a film of dark smoke appears above one of those remote "points"; instantly a negro drayman, famous for his quick eye and prodigious voice, lifts up the cry, "S-t-e-a-m-b-o-a-t a-comin'!" and the scene changes! The town drunkard stirs, the clerks wake up, a furious clatter of drays follows, every house and store pours out a human contribution, and all in a twinkling, the dead town is alive and moving. Drays, carts, men, boys, all go hurrying from many quarters to a common center, the wharf. Assembled there, the people fasten their eyes upon the coming boat as upon a wonder they are seeing for the first time. And the boat *is* rather a handsome sight, too. She is long and sharp and trim and pretty; she has two tall, fancy-topped chimneys with a gilded device of some kind swung between them; a fanciful pilot-house, all glass and "gingerbread," perched on top of the "texas" deck behind them; the paddle-boxes are gorgeous with a picture or with gilded rays above the boat's name; the boiler deck, the hurricane-deck, and the texas deck are fenced and ornamented with clean white railings; there is a flag gallantly flying from the jack-staff; the furnace doors are open and the fires glaring bravely; the upper decks are black with passengers; the captain stands by the big bell, great volumes of the blackest smoke are rolling and tumbling out of the chimneys— a husbanded grandeur created with a bit of pitch-pine just before arriving at the town; the crew are grouped on the forecastle; the broad stage is run far out over the port bow, and an envied deck-hand stands picturesquely on the end of it with a coil of rope in his hand; the pent steam is screaming through the gauge-cocks; the captain lifts his hand, a bell rings, the wheels stop; then they turn back, churning the water to foam, and the steamer is at rest."

Soon this steamer would pull up her gangplank and head downstream for St. Louis and Memphis and Baton Rouge and maybe on to New Orleans. These crowded passengers included fine gentlemen and their wives in the best staterooms, who met socially in the spacious cabins beneath glittering chandeliers to listen to polite conversation and classical music, or to the famous tall tales of the pilot who had a differ-ent story for every bend in the River. Joseph Holt, Ingraham, writing about the river in 1850, tells us:

"You see ma'am, in them old Frenchified times, folks didn't care 'mazing much 'bout law, nor gospel neither. . . . There was a regular band o' pirates lived on the river where Baton Rouge now is, and they had a captain, and numbered fifty men or more—awful rascals; every one of 'em—had done enough murder to hang seven honest Christians. This captain was the essence on 'em, all biled down for deviltry and wickedness; and yet they say he was young, almost a boy, plaguy handsome fellow, with an eye like a woman and a smile like a hyena; and his men were as afraid of him as death. Well, he lived in a sort of castle of his own, over on the little rise you see, near the town, and people said he had (begging your pardon ma'am) as many wives as old Captain Bluebeard, and killed 'em as easy. Well, he had a lookout kept on the point just in the bend and when he had warning of a boat, he'd wind his silver bugle and collect his men to the boats and shouting like twenty heathens, they would dart out with their seven barges upon the descending craft. It

was short work they made then. A rush, and leaping on board, a few pistol shots and cutlass blows, and the crew were dead or overboard. The prize was then towed into the cove beneath the castle and plundered and set on fire. Them were rough and bloody times, Miss!"

Here the canny tale-teller would pause ostensibly to relight his cigar, until the slightly scandalized lady passenger could contain her curiosity no longer and would ask, "Please tell me, what became of this man and his crew?"

"Some say he was shot in the Public Plaza in New Orleans by the Spanish Governor; but I heard an old pilot say that he was assassinated by a young woman he had captured; and that is likely by all accounts. . . . That's the story I hearn, ma'am but I won't vouch for its Bible truth, for it's might hard reckoning up things happening so long ago."

There were lesser passengers, too, and sometimes even suspicious looking chracters who shuffled decks of cards and invited strangers to join in a game.

Up and down the river plied the packets, glittering in the darkness, or shining white in the sun. Stand-ing on deck, a traveler could often glimpse broad fields of cotton, white with bolls, or level green acres of sugar cane; and back from the water's edge the tall pillars of the planter's homes. If the reader of this volume had been such a passenger on a boat bound down the Mississippi for New Orleans in the 1850's, he would have seen the *grand parade* of the architecture of the ante-bellum South; and if an architect had been standing along the guardrail be-side him, the two would surely have agreed that these houses with their gardens and their surround-ing quarters were the fullest contribution of the Old South, both to art and to the art of living.

The writer may be forgiven any seeming partiality, but it appears that a tour of River Houses should start at Memphis, Tennessee. Since 1807, when Cum-ing wrote of "Ascending to Fort Pickering (site of Memphis) by a stair of 120 square logs . . . ," Memphis, on Chickasaw Bluffs, was destined to be among the favored of progress. It is significant that the coming of the steamboat was the signal for developments along the Mississippi above Natchez. Memphis and Vicksburg, with many other river towns, had their beginnings during this period.

It is to be regretted that progress has robbed Memphis of its ante-bellum white-pillared mansions. Even while this volume has been in preparation among the last of these landmarks, such as "The Little Greek House," the Littleton-Pettit house, and the Robertson-Topp house, have been destroyed and there remains of them only historic sketches and a few photographs on file in the Library of Congress (H.A.B.S. Records of 1934).

Down toward Vicksburg the river passes through the great delta country of Mississippi where millions

Vicksburg, Mississippi   THE KLEIN HOME 1850

John Alexander Klein was a prince of commerce by tradition. He was a patron of art and architecture, as shown by his stamp left on the "big house" of Cedar Grove Plantation.

of acres of the world's richest farmlands extend inland for fifty miles to the hill country. These are the lowlands which were accessible only to the rich until the coming of the railroads many years after the War-Between-the-States. While today this country, more than any other part of the South, retains the same basic economic plantation system as in 1860, little is left of its architecture. On a recent trip through the Lake Washington section below Greenville, I found some remaining evidence of ante-bellum life, but no examples worthy of note here. The river and fires have claimed them all.

KLEIN HOUSE

HIGH up from the river front, in Vicksburg, Mississippi, where it lies mellowed in memories of the Deep South, is the Klein Home whose gardens once extended to the very banks of the river and whose domain was Cedar Grove plantation (now beyond the city limits).

Vicksburg, in addition to being an important port and center of industrial and commercial wealth in the Mississippi and Yazoo deltas was, like Natchez, a favorite residence for planters whose estates in the delta regions were too hazardous for the health of their families. They built their town houses, therefore, on the bluffs of the east bank from Vicksburg to West Feliciana.

Built in the late period about the middle of the Nineteenth Century, this home is a monument to a price of commerce, John Alexander Klein, who had come to Vicksburg from Waterford, Virginia in 1836 and who prospered mightily with the growing city. The house is among the first estates we pass as we enter the "lower river country." The Klein House starts, as it were, the *grand parade*.

The architecture displayed here shows considerable influence of the deep South. The veranda shares the end gables as it cuts back into them, very much in the manner of Shadows-on-the-Teche in the Bayou Country. The four pillars, with elongated Doric shafts and good capitals, are true to the lower river transitional feature at the bases where they rest on a pedestal extending to the gound. The top of this pedestal is the floor line of the veranda and is emphasized by the cast-iron railing bordering it. The iron work is an interesting feature blossoming into a riot of lacework in the little summerhouse in the left foreground of the illustration. All this is typical of the Vieux Carré in New Orleans, from whence the inspiration probably came. The one story wing pavilions on each side create a favorable silhouette, and are interestingly treated with bays which are unusually successful in their treatment of the window blinds.

Remnants of the original gardens are still recognizable in the nature of flowers, shrubbery and trees, wrought iron summer house and seats, flower urns of stone and two statues, Summer and Fall, evidence that beauty and practicability once flourished together in the atmosphere of prosperity.

There are seventeen rooms in the house, not counting the bath rooms and cellar space. On the south side of the first floor are located two drawing rooms, the library, the conservatory and the ballroom. On the north side of the first floor are the sitting room, the dining room, breakfast room, kitchen, two bedrooms and bath rooms. When the Kleins lived in the house the entire second floor was used for bedrooms, seven in all.

There are rare pieces of furniture in the characteristic hall which runs entirely through the house. A cathedral chair of walnut, two walnut hatracks (one in each end of the hall), an old blue-cushioned divan and a bronze statue of Hannibal are notable among the furnishings.

Generally speaking, it can be said that the furnishings of the house belong to three periods—the Early American, Louis XV and Napoleonic. Many pieces of the original Klein furniture have been distributed elsewhere but there is enough remaining to express the luxury of the early setting.

On the right of the hall are the exquisite double parlors. Elaborate carvings and mouldings decorate the walls and ceilings. The very handsome pier mirror has as its terminal and side decorations the espagnolette (female bust) ornaments. This mirror is flanked by windows, the cornices of which exactly match that of the mirror, making a unified grouping of what appears to be three cornices moulded completely together as one. In the second drawing room, directly opposite, is a duplicate of the grouping, giving balance and a rich atmosphere for the handsome Louis XV furniture decorated in ormolu, the elegant alabaster urns on the Italian marble mantle and the three panelled mirrored *etagère*. The second drawing room opens into the ballroom. Here was the scene of much gayety and beauty.

The Vicksburg Evening Post has the following to say: "An invitation issued by the Kleins in 1873, and the dance program for the party, recently came to light. It follows:

Vicinity of New Roads, Pointe Coupée Parish, Louisiana

LABATUT, FROM 1750

"As old as tradition," say the Labatut family who live here. When Everiste Bana, of Spanish descent, built Labatut, there was the customary oak grove from the river façade.

Mr. and Mrs. John A. Klein
At Home
Wednesday Evening, May 7, 1873
Nine O'Clock
Fancy Mask   Welcome to Cedar Grove

————

| Quadrille | Waltz | Lancers |
| Waltz | Quadrille | Quadrille |

Quadrille	Polka	Mazourka
Polka	Waltz	Galop
Lancers	Quadrille	Quadrille
Galop	Galop	Galop
Quadrille	Quadrille	Reel
	Varsovienne	

Evidently the quadrille was their most popular dance, but one wonders how the ladies in their bustles and many skirts achieved a "galop."

LABATUT

THE old Labatut House on the west banks in Pointe Coupée is just north of the False River section, directly across the river from St. Francisville and the Felicianas. Like Parlange, just below, it is a typical old French Colonial structure. The appearance of this type so far north is unusual, especially within a stone's throw of the English country. The histories of both houses, however, dates back many years before the coming of the English settlers to the Felicianas, when the French undertook extensive development up as far as the Red River. On the east banks of the Mississippi, they hardly came beyond St. John the Baptist's Parish, where the "German coast" began.

The Labatut place is a house one would rather expect to find farther along the smaller bayous or in the New Orleans garden district. Its scale hardly becomes the life on the Mississippi where the great estates often included *garçonières* larger than it is. There were a few French families of means, however, who never yielded to the influence of the grand manner. The Labatuts were evidently among these, for their descendants still occupy the house and are numerous elsewhere in Pointe Coupée Parish.

The house itself is a perfect architectural gem. The river front has a veranda across the entire façade, the basement floor having six Doric columns, with six wood-turned colonnettes on the main floor. The balustrades are of wood, diamond pattern work so frail that it is a miracle the design has stood intact all these years. The door and window openings are very delicate Georgian detail with transom and sidelights of light muntin work. Those of the main floor are somewhat lighter and finer in detail than the basement. The basement is brick, stuccoed, and the main floor is frame.

The plantation front shown in the illustration, which is opposite the river front, is supposed to be the secondary façade. It is, however, a distinctive type most charming in its expression of the plan as well as in the detail which continues that of the river front.

The plan of this house displays a small layout of the early type, three rooms wide and one and one-half deep, the middle room back of the center living room giving way to a service porch which houses the stairway as well as an out-of-door living room. This type of plan was frequently used where ante-rooms were desired to serve as dressing rooms, personal servants' rooms, or nurseries. Often there appeared two large rooms on either side of the center hall instead of one, but in all cases this façade treatment was used.

The stucco at Labatut is still pale yellow, but very little is left of the other original colors of the exterior. The woodwork of course was white; the original roof was probably slates of browns and purples.

The Labatut house, in my opinion, dates from the late eighteenth or early nineteenth century as similar Georgian details appeared in Natchez and the West Felicianas just across the river at the turn of the century. If these details were not here, I should place its construction thirty years earlier; about the same time as that of its neighbor, Parlange.

New Roads, Pointe Coupée Parish, Louisiana

PARLANGE, FROM EIGHTEENTH CENTURY

We reckon the age of Parlange Plantation House by the Parlange tomb in the New Roads Cemetery which bears the date, 1757. Mr. Walter Parlange still holds forth in the "big house" today.

PARLANGE

POINTE Coupée Parish is a rich region in Louisiana on the West banks of the Mississippi, opposite St. Francisville. When the Mississippi long ago left its channel to form a new one, it created False River and a fertile lowland. This was one of the first up-river sections to be settled by the French because its rich soil encouraged the planters.

Parlange is a true Louisiana Colonial home of the larger type, showing some confusion and signs of additions and changes.

The plan is very similar to that of the Keller plantation house, which is illustrated here. It is two rooms deep and four and five wide, all rooms opening onto the gallery which is the principal communication between rooms as well as floors (by way of the gallery stairs).

The river front, with its two *pigeoniérs* happily placed flanking the entrance, is probably the most interesting of its type remaining today. The trees are so tremendous and the moss so gracefully draped, I have often wondered if the designer of these oak alleys could possibly have conceived of their beauty of a century and a half later. Alas, that he could not have lived to see his planting fully developed. I have wondered, too, what inspiration the earliest French home builders had for these avenues of oaks. Unless they remembered the gardens of Versailles and Fontainebleau, I can think of no other source of inspiration than that of the forests around them in the New World.

Family tradition has it here that "an exquisite formal garden was originally conceived after the *Jardin des Fleurs* at Versailles and carefully tended by French gardeners brought over for that purpose and it was destroyed by the War-between-the-States, when it was used as a mule lot." In all events, we may be sure these efforts served to inspire later plantation settings like those at Oak Alley and Uncle Sam.

The Marquis de Ternant came to Louisiana from France in the 1700's when Baton Rouge was just a small village. He obtained a grant of land from the King of France on beautiful False River. The second Marquis de Ternant in Louisiana, following the democratic trend of the times, renounced his titles and honors and was known simply as Monsieur de Ternant. The present owners of the house, heirs in a direct line, are Mr. and Mrs. Charles Walter Parlange, sixth generation of the family to occupy the house.

The original Parlange remains today nearly intact except for a few changes including the front steps leading outward from the main floor gallery. The present steps were evidently added at a later date because the true Louisiana Colonial type, to which this house belongs, invariably had the stairs housed entirely within the cover of its galleries.

The raised basement is of brick construction and practically a duplicate of the main floor which is of stucco wall construction. The galleries probably extending at one time the entire perimeter of the plan, but somewhat confused now, are supported below by simple undeveloped brick and stuccoed shafts of classical inspiration, while the main floor has the traditional light colonnettes. The floors on the ground level are paved with brick; the main gallery is of wood. The usual simple hipped roof of steep pitch, broken occasionally by one dormer on the short elevations and two on the long elevations, is carried out to the letter. Inside chimneys pierce it at symmetrical intervals and extend high above with a simple brick cap. The roof is of cypress shakes which have weathered to a silvery gray.

White is the predominating color; even the brick walls of the basement floor, deep in the shadows of the galleries are whitewashed and toned down to a dull gray by the cool shade.

On the interior, rich ornamentation is restrained and unusual detail is combined with a very great regard for the beauty of simplicity. The ceilings of cypress boards are so finely joined one can scarcely tell them from plaster. Even the chandelier medallion in the ceiling is of carved wood in the plume pattern of the Empire period. A hand-carved mantel in soft-toned old wood has double columns supporting the shelf; fireplaces are deep and project far into the room. The wallpaper (where used) is not the original hand-blocked, but is of a similar pattern in dull golds and reds of the Empire period, making a pleasing background for the happy union of American Colonial and French Provincial furniture. (At the time Parlange was furnished, there existed an *entente cordiale* between our early republic and the French.) In the parlor, or salon, there is a girandole, giving light and life to the room with its many reflections. Low handcarved French sofas lend a charming note of intimacy in the conversational group before the fire. There are several pre-Revolutionary American pieces of furniture in this home, an early American spinet and one of the first Pleyel pianos ever made. The most beautiful and mellowed touches in the old French interiors are the embroidered tapestries, and in Parlange stands an interesting frame where the ladies of long ago whiled away their time with their petit point. Another characteristic piece of furniture is the Circassian walnut bibliotheque, housing many rare old books; glass

Vicinity of Geismar, Ascencion Parish, Louisiana

BELLE HELENE 1840

Apparently Duncan Kenner appeared in Louisiana in time to join the big parade and build Belle Helene. This old house is still glamorous in spite of its dilapidation and poverty.

panes of the doors of this case are lined in sunburst style with yellow silk.

History teaches us that Napoleon regained Louisiana from Spain long enough to turn it over to the United States in 1803, but Madame Claude Vincent Ternant, being a French woman of fashion, retained the French spirit at Parlange as did she also maintain her salon in Paris until the time of the War-between-the-States. While the war clouds were brewing, her temporary absence from the States gave her a neutral attitude and it was she who saved Parlange at that time, for the Federal soldiers camped all about the place, as did the very welcome Confederate troops at another time. She made friends with the invaders and received them royally, giving them barbecues and other feasts fit for kings. Consequently, Parlange was not ransacked and nothing was taken except one of the children's ponies, and the soldier who took this was made to return it. However, Madame Ternant had taken the precaution of laying the silver along the fan lights where it was out of sight and temptation.

BELLE HELENE

Belle Helene reposes well back from the east banks of the Mississippi about twenty-five miles south of Baton Rouge, near the village of Geismar. Instead of the customary avenue of oaks leading to the river, Belle Helene lies deep in an open green meadow with the giant oaks making a horseshoe setting around its sides and plantation front and then forming avenues to the sugar house and quarters. These old oaks are today the most gigantic to be found along the river, and they are in proportion with the home they surround.

Belle Helene estate was established by Duncan Kenner several years prior to the middle of the nineteenth century. The plan of the house is very similar to that of the Uncle Sam Plantation House (illustrated in this Chapter) with a central great hall or living room running the depth of the plan flanked by rooms on either side.

The inside staircase is housed in an alcove off the rear of this hall, while additional stairs are off the gallery which extends the entire perimeter of the plan. This gallery, twenty feet wide, is paved on the ground floor with beautiful patterns of brick and tile.

The four façades are all alike, each with eight enormous pillars of brick and stucco nearly four feet square and thirty feet high, supporting an exceptionally heavy entablature. The hipped roof is too flat in pitch to show from the ground. A second floor gallery is carried by the pillars at their halfway height and bordered by the customary simple wood balustrade. Fenestration is the simple French Provincial shuttered opening which extends to the floor in every case and renders the rooms as easily approachable from the gallery as from within. The doors at each end of the great hall are surrounded by side and transom lights, and are treated in heavy Greek detail. The exterior walls are stuccoed and marked off in regular Ashlar stone courses. The whole is a color scheme of weathered, light lemon walls, green-blue blinds, and white pillars and entablature.

All in all, Belle Helene is a very noble example of an important type of house built in the Deep South during the period preceding the War-between-the-States. Its prototype is uncertain except for the plan with the germ arrangement of the Greek temple. Its massive proportions would substantiate this temple feeling. Certainly no house ever displayed a more important scale of living.

St. James's Parish, Louisiana

OAK ALLEY 1836

The oaks, said to be two hundred and fifty years old, date from a former house which was removed when Alexander Roman built this mansion. Here is one of the rare occasions where house and alleys both remain. George Swainey was the architect.

OAK ALLEY

OAK ALLEY, the home of Alexander Roman, a famous Creole governor of Louisiana, was built in 1836 on the west bank of the river some fifty miles up from New Orleans. It is well set at the end of an alley of twenty-eight giant live oaks, whence it derives its name. This alley and the house are one of the few such combinations remaining intact today. In most cases, either the alley has been engulfed by the river, or the house has been burned, torn down, or, if allowed to remain vacant, carried away piece by piece for firewood. Here each oak has a spread of approximately one hundred and fifty feet, and the alley seems half a mile long. The mansion appears to be only a small place when it is first viewed from the river road; not until one gets within four oaks of the house is it possible to outline its mass.

The trees are estimated to be more than two hundred and fifty years old—the vision of a pioneer whose name has been two centuries forgotten. He made his way up the Mississippi River during the latter 1600's to a point near the present town of Vacherie where he cleared a plantation and planted an avenue of evergreen oaks. From this planting grew the trees of Oak Alley, a site prepared at least one hundred and fifty years previous to the mansion, as if with a view of this very construction. Twenty-eight great columns, twenty-eight ancient trees, and twenty-eight slave cabins. Twenty-eight was the magic number of Monsieur Roman, the builder.

The house itself is typical of the temple style, with its twenty-eight columns supporting the galleries around the entire perimeter, very much in the same manner as at Three Oaks.

George Swainey was the architect for this white-pillared house of Beau Sejour, the center plantation of a family principality of three plantations. Such was the original name given by Monsieur Roman to his plantation — the name of his wife's family in France.

Luxury and hospitality were synonymous. It was here that Gottschalk caught the inspiration for his famous Bamboula, by watching the slaves dance. Banana trees with their drooping broad leaves gave atmosphere to this Southern plantation and orange groves flaunted their golden fruit and the cane fields were generous in their yield. And always there were the oaks—a veritable alley of oak trees, that were destined to change the name of this plantation. All went well with the French name, Beau Sejour, until time brought American captains to pilot the boats up the Mississippi; finding the foreign name too hard to handle, they gradually began to point to the giant oaks as "Oak Alley," for the benefit of those on board who might be seeking that destination. So as time passed, the new and popular appellation became the very significant name.

Two notable events of the time were the marriage of Josephine, sister of Jacques and Andre Roman to Valcour Aime's son, and a visit to the three plantations by Louis Phillippe of France.

About the time the plantation became familiarly known as "Oak Alley," the beginning of the end of the ease and hospitality of the Old South left its marks. The happy family principality was shattered —the rich lands of Governor Roman were abandoned, the gardens of Valcour Aime disappeared, the slow decay of "Oak Alley" itself began. It was deserted, behind a wilderness of weeds, with broken sagging roof, delicate railing rent; in dozens of places, one found the result of slow disintegration. Pigs and cattle walked unmolested through the rooms of the ground floor and there was a silence in the old plantation bell that formerly rang out across the wide acres of Oak Alley.

Then, we see the power of man reviving dead beauty, for like a dormant bulb, Oak Alley has pushed its way into sudden bloom again, due to the vision of the present owners and the architect of restoration, Richard Koch. In 1925, Mr. and Mrs. Andrew Stewart decided to buy and recreate Oak Alley. Unfortunately the *garçonnières* flanking the house were destroyed beyond repair, but two small houses at the back, which showed unusual grace of line, were fashioned again: one into a delightful guest house, the other for the manager of the place. The mansion is once more a house not to be forgotten. All has been reconditioned within and without. It has been repaired, redecorated and furnished with antiques suitable to its historic beauty. The balusters on the wide upstairs gallery spread like rare lace in pale green-blue tint matching the decorative old blinds. The woodwork of the doors and windows show the charming rose design anew, the spacious high-ceilinged rooms have been repainted; the open porch, flush with the ground, holds again its herringbone pattern; little Doric columns of the doorway have been made perfect and the belvedere from which Monsieur Roman obtained a view of the swirling Mississippi, or watched his darkies at work in the fields, once more tops the house.

The landscaping is an example of intelligent designing. Hedges of waxy yuccas form boundary lines, flowering shrubs for background and four hundred rose bushes that run the gamut from Louis Phillippe to the fragrant Duchesse de Brabant. Lady Bankshire roses and Confederate jessamine now twine around

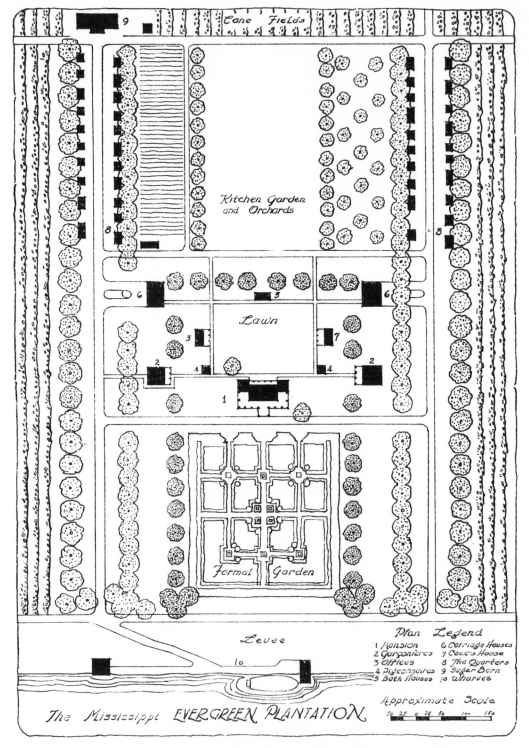

Plan Legend
1 Mansion   6 Carriage Houses
2 Garçonières   7 Cook's House
3 Offices   8 The Quarters
4 Pigeonnaires   9 Sugar Barn
5 Bath Houses   10 Wharves

The Mississippi EVERGREEN PLANTATION

Vicinity of Donaldsonville, St. John The Baptist Parish, Louisiana

EVERGREEN, THE PLANTATION GROUP PLAN

Farm groups are considered first elements in land planning. Early Louisiana plantations, however, were different. Evergreen was comparable to a small principality, incorporating a definite domestic life within an independent agricultural and industrial estate. Ships, some flying foreign flags, docked here for cargoes of sugar in exchange for manufactured articles.

the side columns of the house, orange groves flourish again, crepe myrtles and oleanders border the winding drive. Banana trees, bamboo and feathery parkinsonias edge the approach roads of the plantation; fat cattle roam the pastures and even the "quarters" show the spirit of rehabilitation, each with its diminutive garden.

To view the majestic home in its shadowed pink through the tunnel of green is breath-taking. To enjoy Louisiana architecture, see Oak Alley and Parlange first, then look at Labatut, Belle Helene, Uncle Sam, Three Oaks, Rene Beauregard and all the rest, using your imagination which has been stimulated at Oak Alley.

Tradition has it that after the abandonment of the plantation, there was but one visitor to the place. "Day after day, a carriage would drive up and a frail old lady, meticulously dressed and carrying a parasol of black lace, would descend and motion to the coachman to wait. Under the oaks she would walk, gazing at the forlorn house at the end of the vista. Then, silently, she would go back to the carriage and drive away. Who was she? What memories of youth stirred her tired brain, what regrets, perhaps, pulsed through her heart? Since the restoration of the place, she has never come, yet one cannot help but wish that she might walk once more down the alley of oaks and see the peace, the dignity, the pristine beauty that is again the birthright of Beau Sejour!"

As a writer in *The Progress* has said, "Monsieur Roman yet lives, and will continue to live so long as one column of his mansion remains, or one great tree of the alley rears its branches into the sky. So long will the Old South live. And when these pillars and trees have crumbled, as all things must crumble, the legend will be stronger than the actuality."

EVERGREEN

On the west banks in St. John the Baptist's Parish, just south of Donaldsonville, we find the great sugar estate of Evergreen. Here again is that grandeur of scale found in the Uncle Sam plantation group which is so expressive of the sugar cane country. While it falls short in scale as compared with some of the great estates, Evergreen, in its prime, was second to none in beauty and the completeness of the plan layout to serve its important occupants. Refer to this reconstructed plan while reading the chapter, because what one finds today is but a ghost of the former glory of Evergreen.

Most noticeable and interesting is its grand alley leading back from the river, typical but original in its conception and diversity of planting. The alleys at Uncle Sam plantation are arranged in three, each with a single row of oaks on each side and each terminating in an important pillared building; an additional outside alley on each side borders the entire formal group, runs its length, and separates it from the spreading fields of cane. The group at Jefferson College is flanked by a double oak alley on each side framing the great pillared buildings and leaving the center to a large formal garden. Evergreen combines these two arrangements. Here, there are three rows of trees forming two alleys that flank each side and frame an enormous formal park between the river and the house groups. The outer row is giant live-oaks, the next row is magnolias slightly smaller in scale, while the third or inner row is of dwarf cedars still smaller. Similarly the taller garden shrubs border the alleys and taper down to the small formal blossoms of the flower beds which form a blanket of colors while silvery shelled paths divide it in patterns.

The whole, seen from the river, forms an arena of myriad colors at the end of which proudly reposes the simple white-pillared great house.

While all this formality prevails at the river front, the outermost avenues formed by the oaks and magnolias pass silently by the great house and continue to the "street" and to the sugar house. The second avenue formed by magnolias and cedars pauses however, in passing, to frame another beautiful little formal garden and *garçonnières* which ties in with the *pigeonniers*, offices, and kitchens to decorate the plantation front. Today the long avenues have felt the heavy toll which time has taken of them, but the remaining oaks and magnolias and cedars make up in picturesqueness what they lack in numbers.

The great house is a very interesting later Louisiana type. It combines the traditional plan and hipped roof with the balustraded roof decks and pediments of the upper country. The order remains simple, light Doric with the shafts supporting a main floor gallery. Most significant, however, is the wood, monumental, free-standing staircase which leads up to the pediment entrance via one grand gesture of a "U" curve. The stairway off the gallery on the plantation front is equally graceful, both displaying a remarkable skill in stair building, particularly in the carving of the hand rails.

Evergreen dates from the first quarter of the nineteenth century, but the offices and *pigeonniers* show much earlier construction details. The plantation house is deserted now, although the owner lives up the road in a modest little cottage. The "Big House" is too big.

Vicinity of Hahnville, St. Charles Parish, Louisiana

KELLER PLANTATION, LATE EIGHTEENTH CENTURY

Everything here is Louisiana (Colonial). The veranda staircase, so indigenous to Louisiana houses, has reached its height of architectural development in the Keller House, built by the Fortier family in the eighteenth century.

KELLER PLANTATION

On the west banks at Hahnville, in St. Charles Parish, is the plantation home of the Kellers, built by the Fortier family in the last quarter of the eighteenth century, and for sixty years occupied by the family of Mrs. P. A. Keller. It is a true Louisiana French Colonial house of the largest type; and, along with Parlange, is among the few left today. The house has not only the same plan as Parlange, but so many of the same peculiarities and details that one is forced to conclude that they were erected by the same master builder at approximately the same date, certainly during the same period.

The great house was originally accompanied by the usual *garçonnières*, *pigeonniers*, carriage house and other detached parts of a large scale plantation. The

housed in *garçonnières* along with the boys of the family.

Special attention is called in the Keller house to the variation in room sizes. Some are small, some large, some larger—with no particular reference to their location or importance. They are placed irregularly over the plan, but invariably lead out onto the gallery, some with one opening, some with two or three.

The interiors are simple. There are no carved marble mantels or plaster moldings and cornices such as are usually found in the later houses of the nineteenth century. The woodwork is sparingly used at the base and casings, but is consistently in good taste. The floors are wide cypress boards, and the mantel-

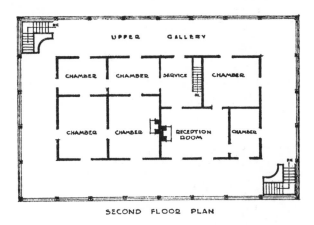

THE KELLAR PLANTATION HOUSE

detailed plan here is of the house only, but it will serve to study the type. There are the usual two floors, the ground or basement floor, built on the ground level and of brick, then the main or second floor, built, as nearly always, of wood. The room arrangement is two rooms deep and four broad, all opening onto the continuous galleries. At diagonal corners are found the two gallery stairs one of which is illustrated here. There is a small sneak stair from one of the rooms on the main floor to the service room which flanks the dining room below, also continuing to the attic, but clearly an emergency measure. The main communication between floors is from the galleries, via these outside stairways.

The basement rooms housed principally the wine cellars, dining rooms and their service rooms, in which was received the food prepared in the detached kitchen fifty feet from the house. There were several rooms for use by personal servants and sometimes by white employees who came on special missions to the plantation, although most of the time, the latter were

pieces are also of wood. The side walls of plaster are covered with the original paper which, along with the draperies, furnishings, floor tiles, marbles, and roofing slates, were imported from France. The wine cellars were extensive, with racks for many thousand bottles and great platforms for the barrels and hogsheads. Marble floors of alternating green and white blocks express the importance of this cellar department, which opened directly onto the gallery as well as into the main dining room.

The galleries on the main living floor are of wood with exposed beams; the floor is wide planking. The stairs are very effective, and the designer took particular care in using a corner to demonstrate their delicate construction and fine carved handrails and balusters—all of wood. Elevation features are the same as at Parlange. The brick and stucco pillars supporting the main floor gallery have the same unfinished capitals and bases, while the simple, light cornices and gallery balustrades are identical in construction. The original roof is now covered with

Lutcher, Louisiana

PRIEST HOUSE, JEFFERSON COLLEGE 1830

This is the gold coast and here was Jefferson College, Valcour Aime's gift to the Marist Fathers and the youth of Louisiana. The Greek revival influence is nearing New Orleans.

metal, but remains in place.

A delightful custom of Louisiana is coffee drinking, and often when visiting these plantations one is invited in to drink a cup of coffee and chat with the hosts. On one such occasion when I asked the reason for all the old enormous cast iron kettles strewn over the State of Louisiana in the Cane Belt, the answer was "Sugar." These kettles were used in the original method of manufacture. Mrs. Keller's uncle, who was a very old gentleman in 1932, pioneered at sugar making when a young man, and it was from him I had the story of how sugar was made sixty years ago.

First, the cane was crushed by large cast iron rollers, which were first operated by oxen, driven in a circle (very much as brick was made). After 1830 most of the cane was crushed by power of steam. This juice was drawn into vats. From the vats, it was dipped into the evaporator. The evaporator consisted of a large bricked furnace, in the top of which was embedded five large cast-iron kettles, graduating in sizes from the largest—twelve ft. in diameter, to the smallest three ft. The smallest kettle was nearest the fire box, the largest at the other end, near the chimney.

The cane juice was first poured into the largest kettle, where it simmered. At certain intervals, it was transferred to the next kettle by means of a large wooden bucket on the end of a cantilevered pole. As scum formed it was dipped into the scum gutters which run the entire length of the evaporator and upon settling, was reused.

When the juice reached the smallest kettle, it was in the form of a thick syrup, practically free of all water and ready to crystalize. It was then removed and put in large containers, where the crystalization was completed. The juice that dripped from it, was the original sugar house molasses; that which remained, was the pure brown sugar.

The Keller Plantation, as well as a great portion of Louisiana, was conceived, constructed and has existed for over one hundred and twenty-five years by the sugar industry.

JEFFERSON COLLEGE

As we pass along the "German coast" near the quaint little town of Lutcher, the east bank suddenly bursts forth with a parade of white pillars as though the entire parish of sugar planters had grouped their mansions on a single plot. In the distant background there are enormous central clusters of columns, flanked by lesser groups on each side. Then the march starts forward, and two great arms of columns reach out toward the river. At the river's edge, there is a high iron fence with imposing gates which invite one immediately into the two charming little gate houses, each with its own columned front. The central motif from the gate houses back to the column parade is a riot of blossoms, all arranged in ultra-formal plots and designs. The whole is framed by twin alleys of live oaks on each side, oaks a century old with great moss-laden branches that touch the ground and then crook upward again, as though resting on their elbows. The visitor is astounded at such a spectacle, such dignity and beauty. Surely a shrine could not be buried in this country and escape the attention of the public for a century; but Jeffer-on College has done so. It was built in 1830, a gift from Valcour Aimé, a rich sugar planter, to the Marist Fathers to aid them in educating the young men of Louisiana. Until 1928, it continued to discharge this duty; now the buildings stand empty.

On the extreme left, terminating the left inner avenue of oaks at the river's edge, is the priests' house. While the pillars of the college buildings are heavy and strong enough to carry their load with solemn responsibility, those of the priests' house, on the other hand, play at their task. Six tall graceful Doric shafts stand twenty diameters high to the entablature as though they were laughing at the heavier columns and their classic nonsense. The usual second floor gallery breaks in between at the midway line, and is bordered by a simple wood railing. The light wood entablature crowns the columns, continuing the same light feeling as it makes its way around the entire house. The walls are of deep salmon brick and the fenestration is the usual simple square top with blinds. The central doors, on both the first and second floors, have charming full-arched fan tops and side lights with all the grace of Georgian detail one could imagine. In plan the house is pure early Louisiana French Provincial, one room deep and three wide. The central doors of which we have just spoken carry one into a central living room from which a stair leads to the second floor. The rooms on each side lead into the hall as well as directly onto the gallery.

The queen of the Jefferson College group, however, rules from a secluded corner in the background, reached more easily through the east gardens and clothed in a seclusion in keeping with the reverence she is entitled to—a little chapel. Small, dignified and graceful is this gem of late English Gothic and

Lutcher, Louisiana

JEFFERSON COLLEGE, THE CHAPEL 1830

Deep in one corner of the garden of the Jefferson College group, surrounded by giant live oaks and pagan temples, will be found a graceful little Gothic building. You must see this—the lovely Chapel of Old Jefferson College.

how wisely its importance has been emphasized here amid this outburst of pagan architecture. The little chapel seems to fulfill its responsibility in a manner different from the priests house; in a manner displaying however, all confidence in its own function as the focal point of the culture fostered by its priests and absorbed by its lay inmates.

The well-formed buttress finials of Lady Chapel reach skyward above the roof parapets to meet the moss of the majestic oaks. Together in the shadows, they invite one to sit for hours in their peaceful presence and search his life for good deeds. Who could leave this spot without carrying away with him a better understanding of the life and culture of the Deep South and of how its young men were molded for the responsibilities of preserving its heritage? Here the Gothic tradition of Christian architecture has found a place amid the level green fields and tall trees of the great Mississippi Valley.

UNCLE SAM

NEAR Convent on the east banks is the Uncle Sam Plantation, which has the most imposing group of buildings of all of the early places along the river. It also has the most elusive history. According to local tradition, it has "just always been there and had that peculiar name." Many theories are advanced concerning both name and origin, but since none seems very credible, we shall derive its history from its architecture.

Although the river has claimed the oak alleys and other important features, there is enough left (1939) to reconstruct a good working plan of the estate as it was in the great days. The whole plot, located by a cluster of oaks towering high above the green fields on three sides and the silvery expanse of water on the fourth, strongly resembles an oasis. The two oak alleys, one on each extremity of the formal development, once reached from the river docks back to the end of the "row" and the sugar house and barns; where they ended the cane fields began. In between these two alleys, the great house, of course, occupied the central position, and every other feature radiated symmetrically from it according to its relative importance. On each side, facing the river and occupying the secondary place of importance, was a garçonnière. These three white-pillared buildings, each at the end of its alley of oaks, formed the first line of grandeur for the benefit of the river approach. After this, the two great pigeonniers and two offices formed a second line of interest. They fronted the east side of a cross avenue, running from alley to alley; then the formal plots and orchards extended on back to a third avenue connecting the sugar house and the plantation barns.

The whole architectural scheme of things here was a perfect combination of a domestic estate with a business establishment, all carried out at the same time and with apparently no conflicting interests. There are many such estates on both banks of the lower Mississippi, very similarly planned. The astonishing fact revealed in such an institution as Uncle Sam Plantation is its apparent independence of the surrounding country. The cane was brought in from the fields to the sugar house, where it was crushed and processed into sugar and molasses. It then passed down the two bordering alleys to the docks, where vessels from foreign ports awaited to trade direct with the outside world. Likewise goods from outside were imported into these little principalities through their own wharves. Back in the slave quarters the negroes made their own boots, clothes, and other necessities, and raised all the staple food stuffs consumed on the place.

At the time this was going on, the workers in the inner circle, the domestic part of the establishment, carried on a regular and normal social life. The great house was used by the immediate family and important guests. The garçonnières were for the boys of the family and for guests who were not of social or personal importance, such as business agents. In the offices the business transactions of the place were conducted. The pigeonniers were, of course, useful as well as ornamental. A great number of the estates had them moved out to the river pike, flanking the big house on both sides. Often the lower floor of the pigeonniers was used for a carriage house.

The order employed on all the houses of the Uncle Sam group is a simple interpretation of the Roman Doric, very near Vignola proportion. Many columns of the same order are used here in a wide variation of features. The great house itself is of the later temple veranda type, while the garçonnières are early French Bayou types, both with simple hipped roofs and dormers. The offices are of the Natchez or Tennessee types with good pediments. The order, however, remains the same proportion throughout all these buildings, welding them into one harmonious group.

The period of construction of Uncle Sam would be between 1840 and 1850. I am unable to find the name of the original builders, but the plantation has been owned by a number of important families of Louisiana.

Vicinity of Lutcher, St. John The Baptist Parish, Louisiana

UNCLE SAM 1840 - 50

We are looking north from the south street, across the domestic group: Left—a garçonnière, the "big house" and the river; right—pigeonniere, office, pigeonniere and plantation.

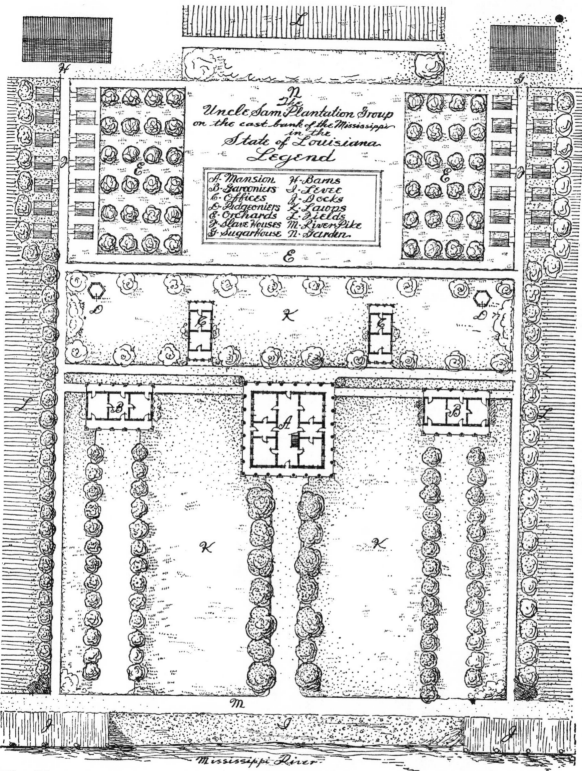

The Plantation Group

UNCLE SAM

Here was a typical Louisiana plantation group of the largest scale. Although the buildings date from about 1840, all plans are definitely of early Louisiana influence and show their French heritage. When one compares farm groups of today with these old Louisiana plantation groups, one wonders why present-day farmers are so careless and unconcerned in their group planning.

Vicinity of Destreban, St. Charles Parish, Louisiana

ORMOND. EIGHTEENTH CENTURY

To the passer-by, Ormond knows no time, owner, architect, or builder. It looks as though it had always been here. Its colors and silhouette soften into the landscape until it is scarcely noticed in passing.

ORMOND

ORMOND Plantation, on the east banks about fifteen miles above New Orleans, is a typical Louisiana dwelling of the four room square plan type with some interesting variations. A study of the plan reveals two long stair halls on each end extending back from the river front the entire depth of the house. From there, two stair halls and galleries . . . The entire

display of colors. One feels a genuine admiration for the house by the time he has reached its entrance. The long expanses of lemon yellow stucco on the central portion are broken by the soft red brick of the first floor of the two end pavillions. The second floor of these pavillions reverts to the yellow, but because it is out from under the shadow of the galleries,

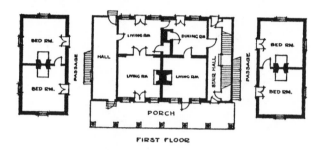

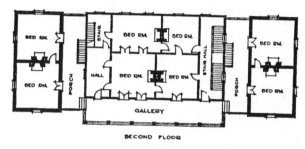

·ORMOND PLANTATION·

plan is covered, but not so happily as if the gallery ran all the way around as in many of the houses we have studied. The stair halls, inclosed, eliminate good exposures, too. Two *garçonnières* seem to have been built so close to the house that in later years it was deemed advisable to connect them to it. That they were separate at one time is evidenced by their decided difference in floor level. The theory of their being originally only one story also presents itself. The plan of Ormond, compared with other early ones, seems to have been tampered with: and I have little doubt that it originally had a gallery on the plantation front as well as on the river front.

The elevation from the river approach is pleasing in spite of its confusing gallery levels and roof treatments. Its great length and symmetry inspire confidence, an effect which is aided materially by the

it has taken on a different hue. From close range the two end pavillion colors are carried across the gallery by the paving brick of the same shade. Pale green blinds and white pillars and colonnettes complete the striking scheme.

In construction, Ormond is typical of Louisiana houses. Its entire central or main portion is built of frame, heavy cypress timbers, mortised, tenoned and pegged together, and filled in with brick and mud and moss. The whole is lathed and stuccoed over. Many small cottages and early pioneer houses were built this way, but for a house of importance, the construction is unusual.

There are many interesting legends connected with Ormond, as with most Louisiana houses. One can secure, in New Orleans principally, excellent books of fiction and history.

RENÉ BEAUREGARD

THE grand parade of Southern homes now passes New Orleans and continues on down the river many miles below. Of the representative houses indigenous to the life in this section, René Beauregard and Three Oaks are especially fine. The former one passes first in proceeding down the river. Built in 1840 by the Marquis de Travis, it is said to have been planned by the elder Gallier, one of New Orleans' first architects and the designer of the old French Opera

House. It was later purchased by Judge René Beauregard, son of the famous Confederate general, from whom it takes its present name.

To me this house is the most interesting example of the transitional period when the capitalist planter was clothing the old Louisiana tradition in a classical façade. Of the few really successful combinations of the two types, René Beauregard is one, probably because of the presence of an architect who under-

Below New Orleans, St. Bernard Parish, Louisiana

RENÉ BEAUREGARD 1840

They "did things" to the Marquis de Travis' house in later years, but we have restored it for him.
Here is how it originally appeared when Architect Gallier the Elder of New Orleans completed it.

stood both the old tradition and the new. The plan shows one of the simplest and earliest French Colonial types, the one immediately following the period of pioneering. Three rooms wide and one deep, it has a veranda on each of the two long sides. All rooms open on to the verandas on both the river and plantation fronts, and at one end is a narrow stair hall which houses a simple staircase (not original). The house as it now stands has acquired two later additions which we have omitted in our illustration because they seem incongruous with the original plan and because they are sadly out of character with the whole place.

Let us see what must have been in the minds of these transition builders as they expanded the early types of houses. To begin with, it was evidently decided that the provincial plan served the purpose satisfactorily, and that to treat it honestly would be in the best taste and produce the best results architecturally. Inasmuch as all the rooms opened onto the gallery alike, the gallery seemed to be an exterior hallway; and as no one part of it was more important than any other, there should be no pediment or special spacing of the columns at any specific point. A simple row of pillars, all spaced equally, was the result. As the same situation existed on the other elevation, the same solution was applied. The order, according to the custom of the day, delighted its patrons when it was used on as big a scale as pos-

sible; consequently, these builders ran it the entire height from ground to roof line and carried the main floor gallery in such a fashion as would not disturb its flight. The entablatures, of course, had to be enlarged from the light cornice of the wood colonette, but at the same time could not appear too heavy for carrying only a light wood frame roof. Because of the very shallow plan, any end gable was discarded in favor of a hipped roof, broken at intervals with dormers and curving slightly as it reached the gallery potion, making in all a very easy treatment. The elongated Doric order applied is graceful. The result was and still is magnificent in its plain solution of an architectural problem so well expressed that the entire plan can be read at one glance.

The color scheme employed at René Beauregard could hardly have been more alluring. The stucco of all the walls is a weathered deep orange, while the blinds, gallery beams, and ceilings are the same madonna blue as at Three Oaks nearby. The slates on the roof range from dark motley buffs to browns and purples; the paving on the ground floor is earthen reds and browns. The columns, entablature, and trim are of course all white. The whole exterior has mellowed quietly with age.

If there ever was a house indigenous to its surroundings, which reflected so simply and truthfully the life of a people, René Beauregard has my nomination for that house.

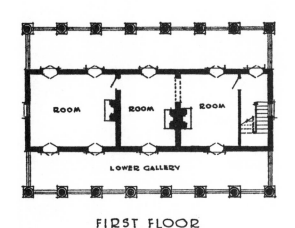

FIRST FLOOR

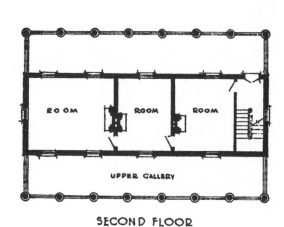

SECOND FLOOR

RENE BEAUREGARD HOUSE

Below New Orleans, Orleans Parish, Louisiana

THREE OAKS 1840

Here the old Louisiana plan tradition holds against all odds, and the palms and tropical entourage of the Deep South assist. I still see more of Louisiana than Classic revival in this picture.

Nor far from René Beauregard, a few miles down the river, is Three Oaks, a mansion of the later ante-bellum period, built about 1840. Here again the plan is true Louisiana, with four principal rooms backed up to each other and all opening onto the galleries. The fireplaces honestly occupy these inner walls, leaving the outside walls to light, ventilation and circulation. Just back of the four rooms are two small service and dressing rooms, flanking a stair hall which houses an insignificant stairway. Basement floors and main floor are duplicates. The galleries running the entire perimeter receive each room via a large shuttered opening and at one corner is a monumental exterior stair done after the best French manner.

The façades have forsaken the old tradition of wood colonnettes superimposed on small columns; one finds the full length, graceful shafts of the classical orders extending from the ground line to the entablature, very much in the same fashion as at Greenwood in the northernmost part of West Feliciana. The main floor gallery here is supported by the column shafts at the halfway mark. The whole construction, including the columns, is brick stuccoed, with horizontal spans of timbers.

The interiors of Three Oaks are simple as in many true plantation homes; the cornices are without ornament and the wood casings and trim fairly light. The fireplaces are treated with heavy wood mantels of Greek detail.

In this particular location the entourage takes on a tropical air. We find the temple veranda type not only applied to the Louisiana plan, but surrounded by a dense setting which gives it an entirely new atmosphere and results in a delightfully different ensemble. The exterior colors are effective. The blinds and ceiling beam work of the galleries are in the most luminous blue I have ever seen which must have originally been like the madonna blue of old Mexico. The stucco is all of light yellow, and the pillars and wood trim white. The hipped roof is covered with slates of soft shades of purple and brown, broken occasionally by a white dormer.

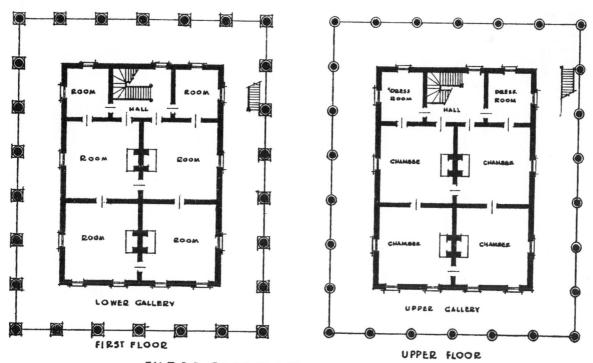

THREE OAKS PLANTATION HOUSE

New Iberia, Louisiana

SHADOWS - ON - THE - TECHE 1830

Here is the river façade of Shadows-on-the-Teche. What style is the house? Who was the architect? Did this façade once boast of a colonnade? Were its deep panelled doorways and excellent wood cornice the product of local craftsmen; and if so, what was the source of their knowledge?

CHAPTER VIII

For Those Who Would Delve Further Into Technical Data

Architectural Styles

As to Architecture and Architects

Available Documents and Guides For Builders

Craftsmanship and Craftsmen

Steam Planing Mills

Brick Making and Masonry

Carpentry

The Portico

Staircases

For Those Who Would Delve Further Into Technical Data

DURING years of research, seeking out old landmarks, photographing, sketching, and leisurely visiting with the residents of old houses, I have always been conscious of their deep appreciation for the physical characteristics of their homesteads. Daughter, mother and grandmother recall the social functions of their forefathers, displaying laces and fans, turning back the edges of the Aubusson rug, comparing the reverse curves of the Louis XVI chair with those pictures in the textbooks, and chatting about keepsakes and trivia. Some member of the family will invariably whisper (as though it were an unmentionable secret), "You should come with me to the attic where you will find not a nail in the roof structure—all mortised and pegged timbers," or, "The brick were made on the place by slaves and the old clay pit is still back of the barn lot." The entire family will agree that the house is well constructed because great-grandfather would not have permitted anything else! Their enthusiasm often waxes so eloquent over technical subjects that they repeat cherished legends, which are incredible to the architect: "There was a fish pond on the roof," or, "Windsor Castle was so named because it was an exact replica of the castle of same name in England," or, "The great Corinthian capitals are of solid bronze."

Yet when this interest in technical knowledge is present and when all of the legends and traditions have been analyzed, corrected, and explained, the owners have a new interest and an additional pride in possession. The rest of us gain the opportunity to use such knowledge in our own work.

Technical data, that is data pertaining to the art and science of building, must of necessity be accurate to be of value. These old houses are often revealing in their design and construction and, when properly analyzed, give up architectural secrets as readily as they reflect the life of the original owners. However, we must know more than what shows on the surface. We must dig into the family heirlooms, old papers, books, contracts, original plans and documents, and contemporary architectural guides. We must seek personalities; we must find out whence came the planners, builders, craftsmen, gardeners, and decorators, and the sources of their learning. We must discover the ways and means of transportation and whence came perfect Greek and classical moldings, elaborate plaster decorations, chandeliers, Georgian doorways, hand-carved capitals, marble mantels, and even artificial gas plants. Who in America designed, manufactured, and furnished the continuous supply of Victorian furniture, and who arranged for elaborate imports from France, England, Spain, and Italy? Such information must be accurate, and that means it must come from only one source—an original document or object. Few of the old private libraries are in existence, although it seems incredible that only seventy-five years have destroyed so much. Attics and yard kitchens and "offices" reluctantly give up some information among the old books and papers mouldering there. One learns that Great Aunt Jennie, during her occupancy as mistress of the "big house," gave Great-Grandfather's "daily journal" to Cousin Ruby who now lives in Peoria; so off we go to Peoria to seek Cousin Ruby. On arrival we find that Cousin Ruby has contributed the heirloom, with other papers, to the local chapter of a patriotic society, the custodian of which is on vacation in Los Angeles! Then there are the various archives of history and art where much documentary information has finally found a resting place. If one is lucky enough to catch up with it, the material is often confined in a glass case or in a rare book room, and birth certificate and good character recommendations are demanded before one is free to study it.

Much has been written about these old Southern houses; but many of the authors, like the residents in the houses, naively assume that all old Southern houses are "good" architecturally. It is surely time to study and evaluate the homes of our forebears as well as to love and preserve them. To those of my readers who want to delve into technical whys and wherefores, to check their own architectural proportions with those of the original Greek and Roman orders, and to form their own opinions regarding the authenticity of the Southern white-pillared houses, I dedicate this chapter.

C H A R T

Showing architectural development in the South (Lower Mississippi River Valley Country) contemporary with that of the Atlantic Seaboard Colonies and the early Republic, and the Old World.

ARCHITECTURE IN THE SOUTH		ARCHITECTURE IN THE AMERICAN COLONIES AND THE EARLY REPUBLIC	ARCHITECTURE IN THE OLD WORLD
1. *Latin influence in the Deep South (through Louisiana, southern Alabama and Mississippi).*	2. *Early Republic influence in the Mid-South (through Tennessee and Kentucky).*		
Exploration period 1699-1750 (French and Spanish explorations and early settlement)		Provincial period 1620-1700 Early Colonial period 1700-1750	ENGLAND *Late Renaissance* Stuart period 1625-1702 SPAIN *Renaissance* Classical period 1556-1650 *Renaissance* Late period 1650-1800 FRANCE *Renaissance* Classical period 1589-1715 Henry IV, Louis XIII, Louis XIV ITALY *Renaissance* 1450-1800 Classical predominating
Colonial period 1750-1800 (French and Spanish colonization and formative period)	Pioneer period 1767-1800	Late Colonial period 1750-1775 Federal period 1775-1820	ENGLAND *Late Renaissance* Georgian period 1702-1830 SPAIN *Renaissance* Late period 1715-1800 FRANCE *Renaissance* Louis XV 1715-1760 Louis XVI 1774-1793 Empire period 1795-1815 Freedom of style 1815-1900 Classical predominating ITALY *Renaissance* Classical continues to predominate
White-pillared period 1800-1861	Glorified Pioneer period 1800-1825 (Formative period) White-pillared period 1825-1861	Greek and Classical Revival periods 1820-1861 Gothic Revival period 1835-1861	ENGLAND *Late Renaissance* Victorian period 1830-1890 SPAIN, FRANCE, ITALY *Late Renaissance* Freedom of style or battle of the styles 1800-1900

NOW-A-DAYS any reference to Southern ante-bellum architecture invariably brings up the question of its style. Greek revival, Georgian, Classical and Gothic revival are terms commonly used. Some offer first one and then the other with assurance that they cover all periods of development. Others are not quite so sure and use the terms with doubt and restraint. Still a third group knows that the South produced a life and culture apart from the rest of America and the world at large, and that its architecture was unmistakably characteristic; consequently these names do not apply, are misleading and uncomplimentary. Obviously it is necessary to conduct a systematic discussion of architectural styles for the purpose of determining the correct term or terms to be applied.

Any discussion of style must, of necessity, begin with a mutual understanding of the meaning of the word. Contrary to popular opinion, an architectural style is a very complex affair. It is not easily defined or understood without a working knowledge of the influences on contemporary life, as well as the history of civilization and art in general.

In previous chapters, Southern architecture has been discussed principally from the standpoint of influences from within. That is, geographical, geological, climatic, social and economic influences. It is now proposed to discuss the prime influences from without; that is, the influence of history and its architectural styles. Such a discussion is best approached in two steps; first, to consider the relation of the Southerner to, and his place in, the evolution of human development. This consideration leads to the modern era where one can chart his architecture in relation to contemporary periods both in America and in the old world. Second, to determine the extent of these influences on his civilization, on his architecture, and in this manner determine this question of style.

Pursuing the first step: the history of architecture is a record of the continuous evolution of human development. Down through the ages from Egyptian civilization to the present, architecture was adapted to meet the changing needs of nations and their development. Whenever a nation has existed within the confines of a given territory, possessed certain soil and geological formations and climate, carried on normal international intercourse and in time developed distinctive economic, social and religious philosophies, that nation has developed an architecture unmistakably characteristic of its national life and culture. The architecture of such a nation is today classified as a style. Obviously all civilizations have had ambitions to blanket the earth with their kind, and to leave behind what they considered a distinctive and original culture. From this desire came the art of architecture. Today, however, it is astonishing to note how few civilizations history has accorded the honor of this achievement without reservation. In fact they can be listed together in one short paragraph.

First, emerging from prehistoric ages, was the ancient Egyptian civilization, dating from 5000 B.C. to 800 B.C. with the contemporary civilizations of Assyrian, China, and Peru. Then came the Greek —800 B. C. to 100 A.D. Third, the Roman—100 A.D. to 400 A.D. Following the fall of Rome came the so called Dark Ages, out of which emerged the medieval periods—Byzantine, Saracenic and Romanesque, which while never fully developed, led up to the final Gothic era of 1100 A.D. to 1500 A.D. Fifth and last, was the Renaissance—1500 A.D. to 1800 A.D.

Two generalities may well be noted. First: these civilizations in all instances had their beginnings in one country, but were not necessarily confined to its boundaries throughout to final development. In the cases of Gothic and Renaissance, one finds that they gradually engulfed the entire continent of Europe before they had run their courses. Second: it is most important to keep in mind the time element. In individual cases, a period of hundreds and even thousands of years were required to develop a style, while all together in 7000 years only five fully developed civilizations and corresponding architectural styles have come down to us. Only one of these, the Renaissance, falls within the era of the discovery and development of America. The Renaissance was an era of rebirth, of free thought. It was responsible for the invention of printing, the mariner's compass and the use of gunpowder, as well as the discovery of America. Architecturally, it was considered a break in the orderly evolution of forms. It was the rebirth of the classics—the original classical revival style, which was destined to be the prime influence on American life and architecture, especially in the South.

To show this influence, a chart presents a clear basis of comparison and can be continuously referred to in order to keep the picture always in mind. In referring to the chart, one finds in the extreme left margin Southern architecture, divided into two influences: first, the Latin; and second, the early Republic. Each of these influences is divided into periods or stages of developments, with dates. Opposite each of these stages to the right are the contemporary periods of the early American colonies and republic, with corresponding dates. To the extreme right are found the various old world influences to compare with both the Southern and early Republic periods.

With this background of history in mind, and the chart of architectural development for reference, one can approach the second step: to determine the extent of these outside influences on Southern architecture. If it can be established that an historic style maintained its form, materials and functional qualities, and in so doing exerted sufficient influence to outweigh the inside influences, then Southern architecture is of that style and should assume its name. On the other hand, if an historic style is recognized only in detail, has lost its material and functional qualities, is clearly dominated by the plan arrangement and materials of the South, then Southern architecture is not of that style and should not be known by its name.

Let us examine these historic style influences one by one.

Southern Colonial: The term Southern Colonial is frequently applied to the white-pillared house of the South by those who claim for it an original, developed style. This is unquestionably a misnomer. Colonial architecture would be the architecture of a colony, if that colony as such developed an architectural style. Kentucky, Tennessee, Mississippi, and Alabama were never colonies. The territory which they occupied was never a colony (with the possible exception of Mobile and a narrow strip of the Gulf Coast). The Carolinas, Virginia, and Georgia were colonies. If they had developed an architectural style during colonial days, it could properly be called Southern Colonial. Louisiana and the Gulf Coast were at times a colony of both France and Spain and developed a colonial architecture (1750-1800). This colonial period was contemporary with the pioneer period of Tennessee and Kentucky, with the colonial and Federal periods of the Atlantic seaboard, the Louis XVI and Empire periods of French Renaissance, and with the late Renaissance of Spain. (See Chart.) It is proper to refer to early Louisiana and Gulf Coast houses built during the latter period of the eighteenth century as French or Spanish colonial, whichever the case may be. Such houses as Parlange, the Old Schertz Home, the Keller Plantation and Ormond, all of Louisiana, are good examples of this French late colonial period. There are no remaining examples of Spanish colonial houses which are of value today, although Concord at Natchez (the illustration is a restoration drawing) shows a strong influence. After the turn of the nineteenth century, this period rapidly gave way to the white-pillared houses, which by 1830 had changed decidedly most of the colonial façade forms. There were no houses in the South built during the nineteenth century which could be called Colonial.

Georgian: The chart shows the greatest Georgian period influences in the American colonies between 1700 and 1775. A continuation of this is found in Kentucky and Tennessee during the pioneer period—1767-1800 and extending into the later glorified pioneer period up until 1825. Rock Castle, Cragfont, and Spencer's Choice in Middle Tennessee, and Liberty Hall, Federal Hill, and Clay Hill in Kentucky are examples.

None of these plantation houses were of the full classical period. However, many contemporary houses were later adorned by the Classical porticoes and verandahs and were well adapted to white pillars. An example is Mount Brilliant in Kentucky.

Later, from 1825 to 1860, much Georgian detail such as windows, doors, cabinets, mouldings and mantels found their way into practically every community in the South. As far removed from this influence as was Louisiana, it too joined in keen desire for these influences and today we find (even in the late colonial) some of these Georgian refinements. Examples—Chrétien Point, Labatut and Shadows-on-the-Teche. However, details do not determine a style of architecture, and as the plan, materials of construction and general utilization of orders were definitely products of the Southern way of life (the influences from within), there seems to be no foundation for the argument that these houses are of the Georgian style. The term would be misleading if applied to development in any section of the South after 1825. Hence, white-pillared houses cannot be called Georgian, though it is proper to give the Georgian period influence recognition for any detail contribution.

Greek Revival: The white-pillared houses in the entire lower valley country were contemporary with the Greek and Classical revival periods of the early republic, as well as with the general revival of interest in Renaissance architecture of the old world (1820-1861). See Chart. It is natural for one to include them in the same category. The term Greek revival comes nearer to applying to Southern architecture than any of the group which we have been discussing. However, a close examination of facts reveals a radical departure from the style in most cases, with the possible exception of detail.

From facts and illustrations throughout previous chapters, one is certainly convinced that the Southerner developed a characteristic plan. His plantation, plantation housing, and big house plan were unmistakably products of his way of living. We are also cognizant of the fact that his materials of construction were definitely local and his craftsmen were trained to know and respect the use and limitations of these materials. The climate taught him the value of shade, and his great verandahs were the resultant original motif and his crowning achievement. During the formative periods of Southern architecture (see chart), his house façades took on forms reflecting these plans, materials and craftsmanship.

It is significant that at this very crucial moment (about 1820) the apparent opportunities in the Southland attracted men of great wealth, intelligence and influence. Accompanying architects brought with them forms fresh from the old world, and soon fine examples of classical monuments sprung up in the form of state houses and public buildings. This was the introduction of European influences as well as those of our Atlantic seaboard. The Southerner at once recognized in the orders of ancient Greece and Rome: columns to support his verandahs, entablatures to span spaces and support roofs, refined mouldings to frame his fenestration, crown his mantels and glorify his interiors. As if by magic, his crude forms took on refinement and his white-pillared house was rapidly developed.

The orderly evolution of forms is an honorable process in architecture. Civilizations are permitted to borrow elements from predecessors, but are expected to contribute their bit towards advancement. The indigenous Doric architecture of Greece adorned itself with Ionic elements borrowed from Asia Minor but at the same time developed its own untold refinement, and emerged with the beautiful orders which were adopted by the Romans. The Romans added the vault and arch and in turn passed them on to the Byzantine, and so on down through the Renaissance to the Southerner. The Southerner, without a doubt, contributed substantially to this evolution. He was most original in his functional use of the orders. He applied them in his own materials and adjusted their parts to function properly and to appeal aesthetically under their use. The mouldings and lesser forms of the classical styles he applied in wood and plaster to suit his own requirements, and emerged with forms definitely different.

I often wonder how close a design can approach the original style without assuming its name. Strickland, in planning the Tennessee state capital building, frankly set out his intentions before the building committee, of using the Erechtheum of the Acropolis at Athens as a pattern. This was satisfactory to the committee, and when they insisted on a tower he simply applied a likeness of the famous Choregic monument of Lysicrates. Here the architect not only used the design, but executed it in the same materials and very much in the same functional manner as the original. Without fear of contradiction, one could call the Tennessee state capitol building Greek revival style. On the other hand, many white-pillared houses are unquestionably remindful of these Greek classical monuments. Some examples are D'Evereux at Natchez, the Priests House, Jefferson College, the Forks of Cypress, Launderdale County, Alabama, and a majority of illustrations in this book. An analysis, however, inevitably reveals that the Greek influence consists only of a few details of orders and mouldings, while the plan and materials of construction and functional qualities are purely Southern. Since it is admitted that detail alone does not determine a style, I can see no reason by these architectural efforts of the Southerner should emphatically and without restraint bear the label Greek revival style. They are admittedly possessed of a strong Greek classic influence.

Gothic Revival: During the years 1835-1861, there was sprinkled over the entire Southland, a decided Gothic influence. Such houses as Ingleside and Loudon in Kentucky, Afton Villa in Louisiana, and Cedarhurst in Holly Springs, Mississippi, are examples of Gothic influence without any attempted use of white pillars. In most instances, Gothic forms were executed in wood, iron, plaster and, in fact, anything except stone—the material of the Gothic style. Very often Gothic towers were used in connection with white pillars, such as on the Walter Place in Holly Springs, Mississippi. There are also times where an attempt was made to use the Gothic style in form of white pillars, such as the Fort House in Columbus, Mississippi, and many others in that city and its vicinity. Here again, however, as in the cases of other styles discussed, the historic style offers little more than the elements of form and detail. At best I think any claim to the Gothic style should be confined to a strong influence rather than to imply that a house is definitely of Gothic style.

Italianate: Among the last of the old world style revivals to sweep the South was the so-called "Italianate." This influence was in reality an adaption of the Late Italian Renaissance as found in the Tuscan, Venetian and Roman provinces of Italy.

Many books appearing about 1850 featured Italian villas. There were bracketed cornices, wooden jigsawn tracery for porches and interior archways, round-head windows, superimposed orders with arches and vaulted vestibules behind them, glassed-in cupolas and towers of wide variety, and roofs flat enough to be hidden behind balustraded cornices.

The Renaissance villas of Italy were of stone, marble and stucco, with tile roofs; but the South had none of these materials. These new forms were irresistible to the Southerner, however, and soon his favorite materials of wood, brick and plaster began to assume different shapes. Among the most daring was the scoring of wood to represent stone and the decoration of plaster to simulate all types of Italian marbles. With the acknowledged limited skill of craftsmanship available, and the short time in which to change these new uses for materials, one can well imagine that the results were anything but Italian Renaissance.

Margaret Johnston's house — Annandale — on the Natchez Trace was a good example of the Italian Villa craze. It was constructed after the design by Minard Lafever, being his Villa No. 4 in the 1856 book bearing his name.

Having discouraged, one by one, the use of historic style names frequently associated with Southern archtecture, it is apparent that one should expect to find an answer within the confines of the South itself. That the South did produce an architecture at least characteristic of its way of life is, I hope, well established here. The one great obstacle in the way

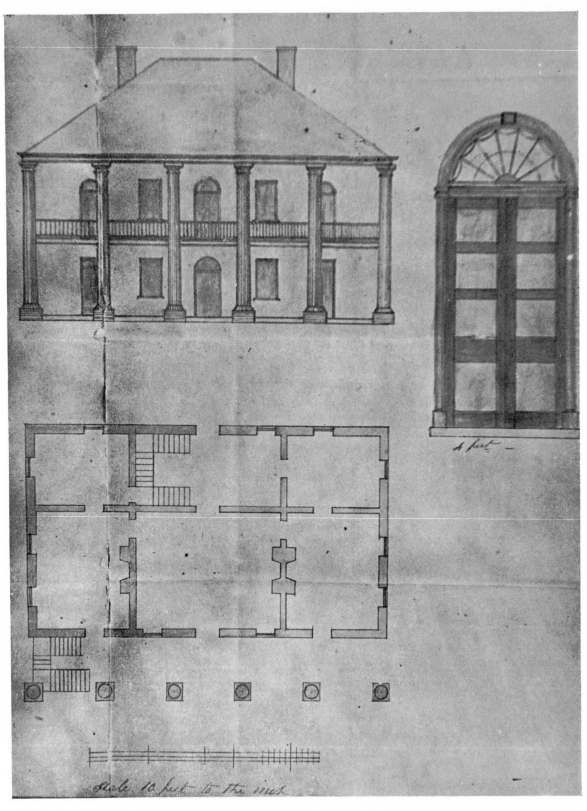

Scale 10 feet to the inch

These are the **plans** of Hypolite Chrétien's plantation house built in 1831 in St. Landry Parish, Louisiana.

of a definite conclusion is the time element. The War-between-the-States put an end to the slavery system upon which was based the South's entire social and economic background. The civilization was never fully developed, but was cut off so abruptly that within four years it was only a memory. Had the South been permitted to continue its particular way of life to full development, probably today it would have developed an architecture unmistakably characteristic, and its style would be known by a specific name. As it turned out, I, for one, choose to forego any style name whatsoever and continue to call the Southern's architectural efforts—white-pillared houses.

AS TO ARCHITECTURE AND ARCHITECTS:

NATURALLY, the first thought when one beholds a stately old white-pillared house, silhouetted against the green background, is of the beautiful picture, and then comes the question of its creator—who was the architect? Local historians usually declare there was no professional designer or add, "Colonel Tucker was his own architect; he designed and built his own house." This theory of the owner-architect is so frequently repeated that it must contain an element of truth. At least it deserves consideration, and provides us with a starting-point in our study of the relation of the architect to the Southern white-pillared house.

Now there was a time in America, not many years back, when all that was necessary for a man to be called an architect was for him to admit it, as the title was not legally conferred until recent years. In the early nineteenth century he seldom used the title; he preferred more detailed terms connected with a particular part of architecture, such as "carpenter," or "surveyor," or even "undertaker." This indicates that the title architect, as we now know it, was of little consequence at the time the Southerner was building his white-pillared houses. Creative and imaginative skill, and knowledge of the building arts and crafts could have been, and often were, vested in the several persons connected with the building. The owner who coördinated these various minds did actually become the master of the situation, and by virtue of enough advice on all sides, became the almost exact equivalent of what is known today as the architect.

Let us take the case of Hypolite Chrétien of St. Landry Parish, Louisiana, who decided in 1831 to erect the good dwelling, Chrétien Point, which we described in an earlier chapter. Here are the plans and specifications which originally served for full contract documents, and from which the house illustrated in Chapter 6 was built:

"This agreement made and entered into between Hypolite Chrétien of the one part, and Samuel Young and Jonathan Harris as Carpenter and Bricklayer and undertakers of building of the other part. All of the Parish of St. Landry and State of Louisiana

WITNESSETH—

that the said Samuel Young and Jonathan Harris undertake and engage to build and construct for the said Hypolite Chrétien a dwelling house near that of his present residence the said house is to be built according to a plan to this agreement anext and made a part of this Contract. The walls shall be brick the foundation sunk two feet below the surface of the earth with end raised nine inches above the ground. All the outside walls in foundation four brick thick and long partition same all cross partitions, two brick. All outside walls above foundation two brick, all partitions, brick and half. The house to be two stories twelve feet each in the clear. The first floor to be a brick pavement gallery—same to be one double chimney above and baloe and one single one same—All the doors in front as well as windows except one door below opposite the stairs shall be circular head as represented in the plan of a door. Five inside doors in lower story to be finished plain with Architrave. All the doors in the upper story to be finished with pannel jambs and pilasters with grounds. All the doors to be double worked pannel shutters. All window frames to be double boxed and each flight of sash to be hung with weights and cord. All windows in the three front rooms of the upper story to be splayed and recessed below to the floor finished pannel jambs with backing and pilasters. All other windows to be splayed jambs, and those in the back rooms in the upper story to be finished with pilasters and all those in the lower story with plain architraves—. All the windows to be inclosed with venetian blines morticed and hung with parliament buts—. Two flights of stairs rampknee. Four handsome well finished chimney pieces and two others. Plain solid water cornice all round the eaves with lead spouts and tin conductors finished, with dentals and modillion blocks. All the rooms and gallerys to be ceiled with plank except the largest room in the upper story which shall be plastered walls and ceiling three coat work and hard finish cornice all round. The front of the wall in the upper gallery to be plastered in the same manner except cornice. The walls of the large room below plastered the same, the Colloms to be well plastered with hard finish. Two small back rooms below to be plastered two coat work and whitened. All other rooms except those as above mentioned to be plastered two coat work and hung with paper of a suitable quality to corrispond with the finishing and stile of the rooms. All the wood work through the whole building shall be painted with a taste and stile suitable to corrispond with the architect and joiners work of the building of the best materials, the roof slate colour. There shall be stone sills to the outside doors. All the locks and fastnings to the doors and windows shall be of the best and a suitable quality to corrispond with the building. The Building shall be completely finished by the Sd Young and Harris or their executors and delivered in fifteen months from and after the date hereoff. the Sd Young and Harris are to furnish all the materials necessary to complete Sd building except the following described —to wit which the Sd Chrétien shall furnish at his own proper expense all the wood necessary such as timber plank shingles and lathes and all the bricks and sand necessary to complete the building and one hundred and fifty tuns of lime the whole on the spot at the building.

In consideration of the work and labour to be done in and about the house and the materials furnished by the Sd Young and Harris the Sd Hypolite Chrétien agrees are promises to pay to the Sd Young and Harris the sum of seven thousand dollars to wit in advance eight hundred dollars when the fundation of the building is laid two hundred dollars.—when the first story is up and joist laid eight hundred dollars, when the building is finished four thousand and two hundred dollars or his notes for that sum bearing ten per ct. in until paid and further the Sd Chrétien for his part shall furnish a house for the Sd Young and Harris and their workmen to live in whilst carrying on the building and shall furnish all the necessary laibourers to attend the workmen in building and shall do all the necessary hauling of materials to Sd building, that is to say bricks, planks, timber, sand and lime.—as—two thunder rodds would be necessary for fear of accidents, the Sd Young and Harris shall furnish them and place them on the building, without any further compensation.

This agreed on, passed and signed before Pierre Laboche, notary public in and for the parish of St. Landry, at his office in the town of Opelousas, in the presence of John Moore and Bernard Vernon witnesses, who have signed with the above named parties on this day the seventeenth of May in the year eighteen hundred and thirty one.

(All the parts of the above written by B. Vernon, as also five words erased, are approved before signing)

WITNESSES	Samuel Young
John Moore	Jonathan Harris
B. Vernon	Hte. Chrétien
Laboche	
Not. Pub.	

Recorded in Book D Pages 87, 88, and 89 of original Acts proper before me in testimony whereof I have hereunto signed and affixed my official seal at Opelousas this 17th May, 1931
Laboche"

The author of these documents is doubtful. He may have been the owner, or Samuel Young, the carpenter, or Jonathan Harris, the brick mason. The drawings could have been furnished by the steam planing mill in New Orleans. That point is not important, because the plans were not the prime factor in obtaining the result—the finished house. Not with the wildest imagination can we believe that a contractor then or today could produce this house from these documents alone. No, the most important factor in this case was that all the parties to the contract had the same house in mind. Indeed, they knew only one kind of house. The carpenter and the mason knew what they were going to build, and the owner knew what he was going to get. They had all been born in houses like that, and had lived in them all their days. Such structures, indigenous to the Bayou Country, belonged to the carpenter, to the bricklayer—to the people. Houses like Hypolite Chrétien's were a part of the civilization rooted in the soil of the upper Bayou Country, and shaped by the habits, loves, hates, and hobbies of the folk. What happened in the cast of Chrétion Point could have happened anywhere in Louisiana until the later nineteenth century period.

Imagine yourself, while building your dream house, being perfectly at ease and dismissing the subject as satisfactorily disposed of, by merely writing a note to the carpenter saying, "hang with paper of suitable quality to correspond with the furnishing and stile of the rooms," and "all the woodwork through the whole building shall be placed with a taste and stile suitable to correspond with the architect and joiner work"— "Install two flights of stairs rampknee" . . . "four handsome well finished chimney pieces"— . . . "all doors in front as well as windows . . . shall be circular head." Without even a mention of the style or order of architecture to be used for colonnade, or materials for the same! Then imagine retiring to your favorite summer resort, or going complacently about your business. That is what Hypolite Chrétien did; and when he returned, he found his house ready, the stairs handsomely done with walnut handrail, balustrade and newels; four Italian marble mantels of excellent design in place; all front doors and windows gorgeous Georgian affairs with deep, rich panelled recesses and perfect fanlight transoms; the wall paper and woodwork in utter harmony. All just as he had known it would be.

These plans for Chrétien Point served merely to describe the parti, while the specifications were a basic contract. Further than that, no written word or drawing was necessary, or could have possibly been produced so as to be understood. These people could no more have pictured or specified that house than they could have described an autumn sunrise in exact colors, or the texture of their handmade brick. Working together sympathetically, the owner and the craftsmen produced a sound, organic house, of which it can really be said that the owner was the architect.

We have been discussing in the case of Chrétien Point and its builders a simple social order, one which evolved slowly from the early French settlements to the American plantation. This house grew out of the early culture and absorbed without loss of dignity the new ideas and details. On the other hand, we find it equally important to contemplate what kind of domestic architecture resulted from more complex changes in Southern social life, involving a violent break with the past.

One such type of change occurred when the Americans from the Atlantic seaboard found a land uninhabited by any other permanent settlers. Their homes, after the first rude shelters, were erected according to the styles of their old homes, and were usually imitative and confused in their architecture. Occasionally they imported an architect or a plan from the cultured East and built a house that was correct according to established conventions but having no essential relation to the new land. Usually it was not until the third or fourth generation prospered that any original or bold idea was projected into their homes.

Another type of development appeared where the American newcomers ignored the already existing native architecture and superimposed, often hastily and ruthlessly, their own houses upon the landscape.

I am thinking in the first instance of the Kentucky Bluegrass region and the Nashville Basin, where, in their formative periods, everyone was still influenced by past civilizations. Some attempted to remember houses of their earlier acquaintance in the Carolinas, Virginia, or Maryland, but being amateurs and without technical guidance in design or proper execution on the part of the various building trades, they made poor contributions. Others wisely enlarged their log cabins into dogtrot houses, added another story, sheds and outhouses, also of logs, and thus created a glorified pioneer ensemble. These plans served their purpose well for two or three generations until their

civilization had become sufficiently rooted to hope for an indigenous type. Such houses as Ashland in Lexington, Liberty Hall in Frankfort, and Cragfont near Gallatin, are examples of imported architectural skill. They are of the late Georgian period, the creations of Latrobe and Jefferson and other well-know American architects of the older civilization of the Atlantic seaboard and abroad. While these houses were fortunate in having such organized minds in architecture, construction, and supervision, others built along traditional lines did not fare so well. By the time Jackson occupied the White House, Tennessee and Kentucky had grown rich and ambitious. Old late Georgian houses took on imposing porticos, and the glorified pioneer logs were finally discarded in favor of white pillars. William Strickland in Nashville and his pupil, Gideon Shyrock in Lexington, designed state houses after the architecture of classical Athens and Rome, as well as numerous extensive homes. With these gifted architects, the new generation established itself in spacious houses. The Upper South developed an orderly type of pillared house somewhat different from those of other sections.

In the second instance, I am thinking of the Bayou Lafourche Country in Louisiana, a sleepy flat prairie land laced with streams, producing a civilization not at all different from Bayou Teche, but unfortunately too close to the "Grand Parade" on the Mississippi, too accessible to the wealth of a world seeking fortunes in vast agricultural estates of cotton and sugar cane to hold fast to its own life. Here the simple indigenous farmhouses which had developed under French and Spanish rule were suddenly routed by great white-pillared mansions of wing pavilion plans, massive in scale. Here the familiar names of early American architects, James Gallier and Son, Charles Dakin, and James M. de Porrily found a fertile field to ply their skill. Between 1825 and 1840 men made fortunes in five years; time was not available for the old and slow process of becoming a part of the soil. With a world demand for cotton and sugar cane, a new civilization could be and was manufactured with great rapidity. Trained architects conceived and created houses suitable to the importance of the plantation princes and in sympathy with the national pride of the New Republic. This was a Classical Revival period, the era of white pillars for Bayou Lafourche, reflecting her own grand manner.

In Mobile and along the Gulf Coast, extending somewhat into the Black Belt of Alabama, as early as 1828 we find Architects Thomas James and Baron von Sleinwelm, laboring with their backs to the wall to execute orders for new houses. Here was a condition entirely different from any in the lower Mississippi valley. Late in starting because of Indian trouble, Alabama, when it did open to immigrants, waited for no one. Architects, carpenters, surveyors, craftsmen—all were in demand; but if they were lacking, owners used any labor at hand to get the work done. When one rides through Alabama today and counts ante-bellum houses by the thousands, of which only a few are good architecturally, one realizes it was a decided case of "no architect." In the decade before the War-between-the-States, the social and economic life settled down and much good building was done.

In all these situations which show a break in the continuity of building traditions, the question as to whether or not there was architectural advice is pertinent. Where conditions favored the use of indigenous plans, as at Chrétien Point, the owner was, as we have seen, often his own architect. In the more complex localities, there were a few imported architects, but frequently the advice of building manuals or the builders' rule-of-thumb practices prevailed until the hey-day of the ante-bellum era when the South developed its own professional architects. We may be fairly sure that few good houses were erected after 1820 without some kind of architectural advice, textbook or professional.

Let us take a look at the American architect of about 1820. In addition to being well fortified with a library containing technical books of the day, he usually possessed university training abroad, and had studied under some established professional man. An example of the best type is Benjamin Latrobe, who was educated at the University of Leipsic in "Knowledge of Every Kind," was a soldier of fortune, an observer of the arts and architecture of Europe, a builder of bridges and canals, a master of all kinds of civil engineering, an architect and an artist, as well as America's outstanding early geologist and mineralologist. He came to America from England in 1796 and moulded a most active career, covering the territory from New York to New Orleans, his works ranging from the design of the United States Capitol and country houses and canals, to the completing of New Orleans' first water works in 1820. He left a wealth of architectural and social sketches, as well as a Journal, which today is probably among our most valuable documents on early American life. He visited socially with Washington and fought furiously with Robert Fulton over steam engine designs.

Men like Latrobe were scarce, not only in America, but in Europe as well. It so happens that the Southwest, as the section was then known, at this very time was a focal spot of activity and naturally attracted men of action. It took men possessed of creative minds and capable of understanding the needs of these people to solve the problems of building the towns and dwellings of the new land. When the man was not available, the outcome was bad architecture, whether waterworks or canals. Even Latrobe, in the end, was the victim of his own inability to think fast enough to keep up with changing conditions. He spent endless nights, lying awake listening to the hum of the mosquito and figuring out canopies of mosquito bars. There were canopies for beds, porches, and offices, even canopies to be held over a gentleman by his slave, as he walked down the street. Latrobe evidently never hit on the idea of screening the houses; and he died of yellow fever, caused by the bite of a mosquito.

Thus Southern architecture developed through the collaboration of craftsmen and owners, and also through the labors of the early professional, who was master of the building arts and of engineering and architecture, and who alone could be depended upon in the confusion of early formative periods. In both cases, it is interesting to study the source of their knowledge as well as its results. Here is the inside story of my technical inquiry.

THE MODERN BUILDERS GUIDE.

BY M. LAFEVER

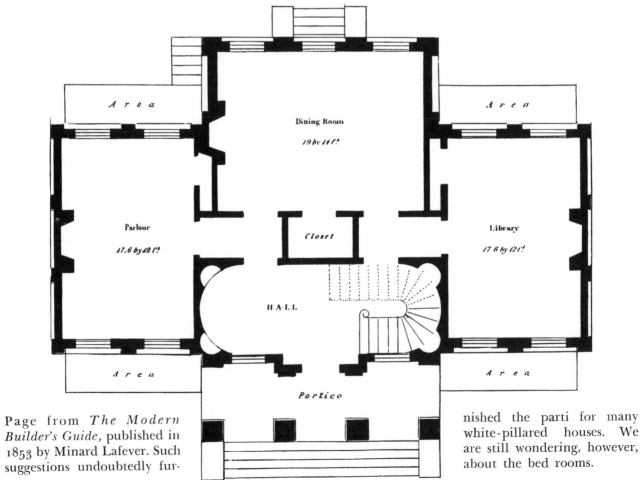

Page from *The Modern Builder's Guide*, published in 1853 by Minard Lafever. Such suggestions undoubtedly fur- nished the parti for many white-pillared houses. We are still wondering, however, about the bed rooms.

AVAILABLE DOCUMENTS AND GUIDES FOR BUILDERS

WHATEVER the actual condition of architectural and structural knowledge in the South during the early nineteenth century, the background was the continuous stream from the printing presses of manuals of architecture and technical information handed down and kept up to date by practical builders. I have selected a number of pages from these old books in order that the reader may study the white-pillared houses from contemporary sources. I have exactly the same book before me as did Gideon Shryock when he designed the stair for the Kentucky State House at Frankfort, or Samuel Young when he constructed the roof trusses for Chrétien Point in Louisiana. The old books are, in many cases, battered volumes, thumbed and dog-eared, which have been preserved on the old plantations.

As a usual thing, the authors of these handbooks did not suggest floor plans and elevations for country houses, but rather confined their efforts to elements of architecture and construction. Occasionally, however, there were several suggested town houses, and once in a while some one would venture an offering in country houses. In 1816, Asher Benjamin in his *American Builder's Companion,* suggested a house "intended for a Country Situation," and Lafever in 1853 added to one of his books designs for country villas. A typical one is reprinted. One will note that the plan is not the simple indigenous one used in any part of the far South, and the elevation is a combination of the architectural elements found in the author's books. At best the design serves to illustrate one way of combining elements into a working whole; undoubtedly this is what Lafever had in mind. Floor plans, he knew, as did his contemporaries, were a result of experience combined with the owner's needs and were not a problem to be solved by any book. Plan types in the new Southern provinces were necessarily different from those in the original Colonial states and often peculiar to particular parts of the South itself. Throughout the first chapters of this volume these types have been discussed freely and none was found to resemble in element Lafever's "Design for a Country Villa." Sloan's *Country Villas,* published in the "fifties," and from which Natchez's Longwood was taken, was one of several publications going the rounds.

The establishment of the building trades of carpentry and masonry appeared as early as the first settlement of any community. At the first sign of transformation from log house to white pillars, these trades acquainted themselves with proper methods of building. Two sources of instruction were available: the acquisition of builder's manuals,

and the emigration of experienced carpenters and masons from England, Scotland, Ireland, and sometimes from Germany. In either case, local tradition, materials, and requirements altered the procedure somewhat; but methods in general followed European standards. In the end such standards prevailed whether brought by the emigrant craftsmen or the builder's guide.

In England and Scotland as early as the turn of the eighteenth century, books began to be published on carpentry, the orders of ancient architecture, and the like. By the latter part of the century, F. Taylor Architectural Library of London circulated in this country a catalogue listing some seventy editions of "modern Books on Architecture—Theoretical, Practical and Ornamental." Nevertheless, Abraham Swan's preface to *The British Architect* declares:

"Many books of Architecture have been wrote, and some of them pieces of great value; yet after careful perusal I have never found any of them fit to give a learner tolerable satisfaction and I might venture to say, there is no book yet extant which contains the rules and examples of drawings and working in so large a variety, and at the same time in so plain and concise a manner, as this single Vol."

Therefore in 1775 an American edition of this book was published in Philadelphia. From then on, the situation of supplying information to the newly created American builders was amply taken care of by American books. In 1786, John Norman published his *Town and Country Builder's Assistant* in Boston, and in 1797, William Pain his *The Practical House Carpenter* in Philadelphia.

Asher Benjamin of Boston published his first book in 1805 and followed with six or seven or others, through 1857. His *American Builder's Companion, or, A New System of Architecture,* published in 1806, carried this statement in its preface:

"Books on architecture are already so numerous that adding to this number may be thought to require some apology; but it is well known to anyone in the least conversant with the purpose of architecture that not more than one-third of the contents of European publications on this subject are of any use to the American artist in directing him in the practical part of his business."

Other American authors of architectural references which were used a great deal throughout the country were Peter Nicholson, Minard Lafever, and Edward Shaw.

Practically all of these early publications started

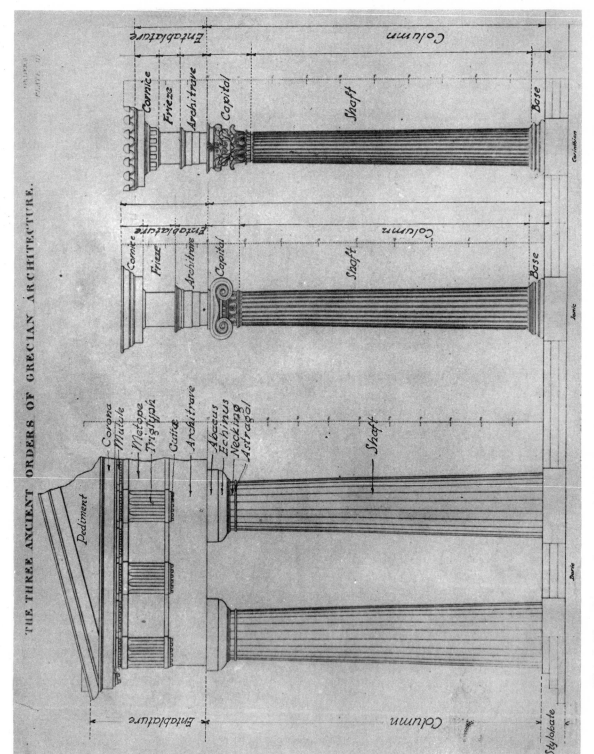

THE THREE ANCIENT ORDERS OF GRECIAN ARCHITECTURE.

This is Plate No. 3 of Peter Nicholson's, *The New and Improved Practical Builder* (1848), showing the "three ancient orders of Greek architecture."

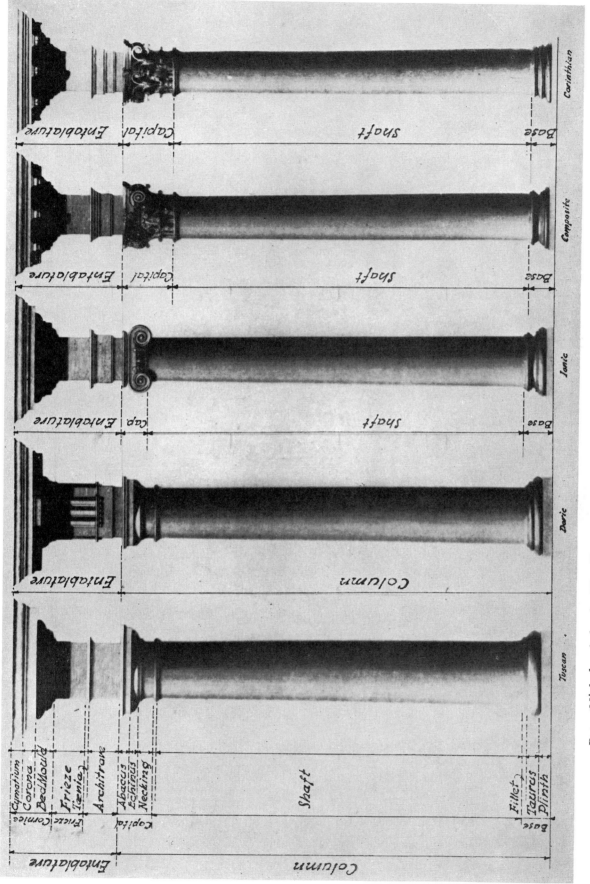

Peter Nicholson's book, *The New and Improved Practical Builder*, published in 1848, shows "the five ancient orders of Roman architecture." It is hoped they will serve now, as then, as examples for those who are too often confused.

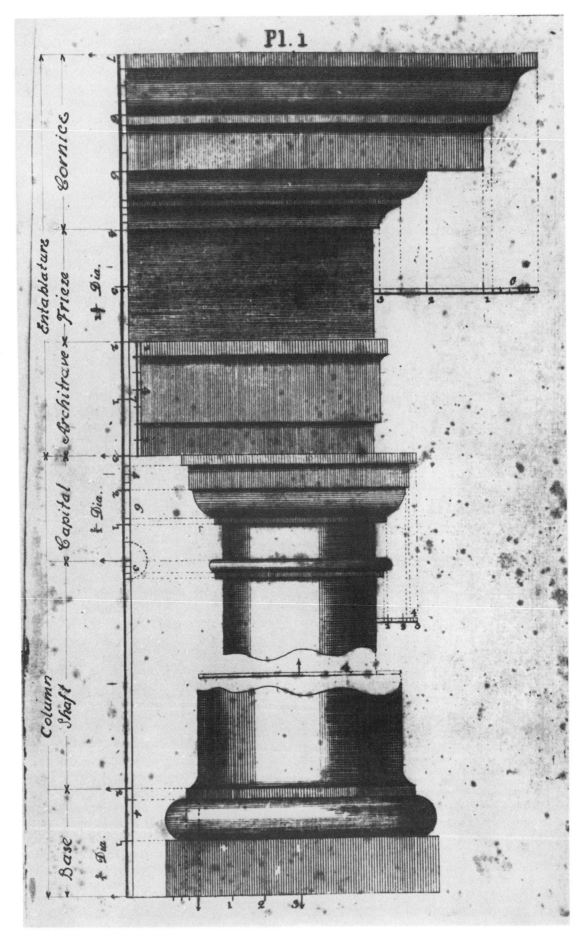

In 1786, John Norman published his *Town and Country Builder's Assistant*. This is Plate No. I, showing the Tuscan Order in all its ancient proportions. Names of elements have been placed by the author.

with a treatise on the ancient architectural orders of the Greeks and Romans: their mouldings, chimney pieces, stairways, doors and windows complete with trim. They took up practical geometry, descriptive geometry, stair building, carpentry, masonry, roof truss design, framing, stair framing, and, finally, specifications in carpentry and masonry. Ample illustrations and drawings formed the second half of the books, which occasionally ended in suggested designs for stores, townhouses, and country "villas."

CRAFTSMANSHIP AND CRAFTSMEN

To be a craftsman one must be an intelligent person, with a natural aptitude for hand-skills. In Europe the craftsmen are apprenticed in childhood and spend years of training until theory and practice are perfected. America's new pioneer civilization had developed no such craftsmen of their own, but suddenly created a demand for the services of trained workers. The building crafts were no occupation for an ordinary person who might carelessly decide to "take it up" in his spare time or at a late age just because the demand existed. Craftsmanship was an art and each branch was a part of a particular civilization. Italians were natural workers in marbles and tiles and glass. The Irish and English from Norfolk and Kent usually excelled in brick. Germans were exceptionally talented in plaster and ornamental stucco, as were the Italians. The Welsh were past masters in brick and stone; Englishmen of Herefordshire and Sussex aspired to be master joiners and stair-builders.

Unquestionably the greatest catastrophe overtaking the architecture of the Southerner was the lack of talent to shape his requirements into form. He had the money, the materials, and the desire for a fitting home. He had the necessary cultural background; but he did not have the necessary architectural and craftsman talent. Fortunately, he did not therefore refuse to build at all. Instead he forged ahead; he would build without talent, and build he did. The results were often crude and clumsy. At the best, four of every five houses standing today show poor design and craftsmanship. Especially is this true in the case of inland communities. Naturally, I have selected good examples for this book, but I make no claim for them as perfect.

The period immediately preceding the American Revolution was the golden age of craftsmanship. The period from the War of American Independence to the War-between-the-States which developed white-pillared houses was the beginning of its downfall. All eyes were on America, where, especially in the rich transmontane South, were men who had wealth and demanded workers to build new communities. But the world which was its direct market could no more supply such a demand than could the South supply it with cotton, tobacco, and sugar. Benjamin Latrobe, surveyor of public buildings for the United States in 1807, remarked that "at this time the country was entirely destitute of artists and even good workmen in branches of Architecture." But the South was not to be deterred. Every known inducement was offered to competent workmen. Colonel Young, at Waverly on the Tombigbee River, entertained his two plasterers from Mobile at hunting and fishing for two months before they "set a mold"; but the job as it stands today, a perfect specimen of ornamental work, was worth the forebearance on the part of Colonel Young.

The South petted and persuaded craftsmen; it offered private homes and fortunes to men of stability and family like Adam Wills, the marble setter of Mobile, Alabama. Romance and adventure attracted the younger generation, such as Henry Miller, a plasterer from Pennsylvania who was working in St. Louis when in 1838 he wrote in his diary:

"Sunday (October 14th) . . . Spent the week in making preparation for going down the river to the southern country to spend the winter there; this is very customary in the upper country; many of the mechanics emigrate from the Southern country in the Spring and return again in the fall of the year. The distance is but 12 or 1550 miles which is brought near home by our steam boats; so much so that many of the young men here take a notion one day and are off the next and think no more of going to Vicksburg . . . Natchez or New Orleans than we formerly did of going but 10 or 15 miles; so much difference do the present facilities of traveling have

on us what the former had. We prepared ourselves for traveling & made choice of the Steam
Boat Alton, Capt. Holland."

These conditions created a very interesting type of
craftsman commonly called the "circuit rider workman," ever on the go, from town to town via the
country estates and then back again. Plantation
owners furnished comfortable quarters and slaves to
help in heavy work and supplied all daily needs;
transportation was always furnished by the next employer. At his own convenience the craftsman would
accept such jobs as fitted into his itinerary and insured
him pleasant weather and surroundings.

There were other types of craftsmen: the master
carpenter or mason, often called the "undertaker of
building," who was permanently located and usually
the employer of local workmen to do common framing and masonry, the semi-rough work. He was seldom
a specialty craftsman, that is, a plasterer, joiner, stair-
builder, or marble worker as was the "circuit rider"
type. Slaves were used everywhere as helpers and
laborers and usually for simple crafts, although the
quality of their workmanship was seldon considered
good enough for the "big house."

It seems that such activities through several generations should have developed craftsmen among the
Southerners themselves. Many of the first of such
people attracted to the South settled there and their
skill passed down to their children, more or less.

However, the temptation to try for agricultural fortunes was usually too great. As soon as any one got
together enough money to buy land, stock, and
negroes on any scale whatsoever, he deserted his voca-
tion to become a planter and a consumer. In fact,
this became so common that it seemed doubtful if
eventually any other occupation than agriculture
would exist. Professional men, school teachers, and
even statesmen bought land the same as craftsmen of
the building trades. The professions and crafts had
constantly to be recruited from newcomers.

In every way he could, then, the ingenious Southerner secured the necessary craftsmen for building.
He offered solid inducements for permanent residence, and he encouraged the "circuit rider"; but,
even so, the demand was greater than the supply.
Workmen in the South could not build fast enough
or big enough to satisfy the prosperous planters until
machinery came to their aid in about 1830. Although
the planters might conceivably have been content
indefinitely with hand methods, the need for quick
results and large structures made them welcome whatever products the steamboats brought them from the
new planing mills. The period of the South's greatest prosperity and influence coincided with the
beginning of the machine age to produce imposing
white-pillared homes as well as the occasional jigsaw
monstrosities.

STEAM PLANING MILLS

WE have seen science develop the steam boat, steam
press, cotton gin, and sugar mill, which increased in
manifold ways the South's resources. Southerners who
were suffering for lack of skilled workmen and were
not able to increase the supply made it possible for
each man to turn out more work by better and improved tools or machines. Thus they participated in
the first "increased production scheme" in the American building industry.

Steam planing mills sprang up in the key cities of
New Orleans, St. Louis, Louisville, and Cincinnati to
supply the entire valley with planed lumber and
moldings, sash and doors, and joiner work of many
kinds. By 1840 each locality had a steam planing mill
of its own. The larger mills increased their com-
modity output to include ensembles of doorways and
window frames of various Georgian and Classical details. All kinds of portico equipment were available,

including complete columns with carved caps, entablatures with dentals and modillions, and mantel
pieces and staircases. All of this came to be known as
the "architect work" or "joiner work" of the house.
The foundation, walls, and roof were the "undertaking" of the master carpenter and mason.

In 1853 in St. Louis, we find this advertisement:

"Missouri Steam Planing Mill
and
Builder's Warerooms
Sash, Door and Blind Factory
Mill and Lumber Yard
Corner of Walnut and Ninth Streets,
St. Louis, Mo.

We have erected a large Steam Mill for the
purpose of manufacturing and keeping on hand
an assortment of Doors, Sash, Blinds, Mantels,

Base Shelving for Stores, Palings for Fences, Weather Boarding, Flooring, and every kind of carpenter work suitable for steamboats and buildings. Boards and planks planed on both sides to any thickness required. Re-splitting, Ripping Scroll and Circular Sawing, Ploughing Relating; also Mouldings of every variety of Pattern prepared at short notice.

Being practical builders, employing none but experienced workmen, we are prepared to furnish work as low as any establishment in the West. Our work is all made of seasoned lumber, and warranted equal to that made by hand. Considering the low rates of freight, carpenters and others about to erect buildings on the lines of railroad running into St. Louis, or on the Missouri or Mississippi rivers, will find it to their interest to purchase all their work and lumber from us. We have a printed bill of prices which we will send to persons who may wish one. All orders will receive prompt attention. Terms, cash.

SAWYER AND McILVAIN"

The mills of course were able to guarantee steady, year-round work which appealed to some of the older craftsmen as they tired of adventure. The mills decreased the number of these men in the rural districts, but greatly increased the output of manufactured material as far as the consumer was concerned. Also, steamboat building had become quite a large item in employment of the wood working crafts.

There is no need to condemn the steam planing mill—it did the job intended and a little more. A mill in Louisville employed designers otherwise inaccessible to Martin Gillard in St. Martinsville, Louisiana. It executed the designs in good workmanlike manner and shipped ready to "set in" rough framework, all the "architect work" for a house, complete windows and doors and porch columns. The result was to furnish the builder a delightful and refreshing choice of styles. For example, in a community where because of one local architect there was a monotonous repetition of Greek detail, the planing mill could furnish a Georgian fan-light or a full semicircular one. Planing mills also maintained large libraries of architectural books from which their designers constantly drew new ideas for the benefit of their customers. Furthermore, the advent of the machine made possible more beautifully turned balusters and newel posts, the latter being very much in evidence throughout the Lower Valley.

Unconsciously, the Southerner did, by so welcoming machined products into the building industry, combine men and machines to increase output per man, and so served actually to render the craftsman less and less important.

BRICK MAKING AND MASONRY

Much has been said about the art value of a handmade brick. The potter never served humanity better than his cousin, the brick maker. The old mason was charged with double duty. He had to master the art of manufacturing his material before he could practice the art of using it. Because of the importance of his job, we often find Master Mason teamed with Master Carpenter as the "undertakers of building."

Brickmaking changed little from the beginning of history until the late nineteenth century, when the railroads made transportation economical enough to haul the brick from a central plant to the building site. Before this innovation, the manufacturing plant instead of the brick was moved to the building site. The movable equipment consisted of two oxcarts with oxen, one mule, and several "moulds." More important than equipment was a place to put it. Such a location, of necessity, must be in close proximity to the building site; and at the same time there must be an ample supply of good brick clay and water and wood for burning—in other words, all raw materials. With this equipment, plus a few pieces of iron, the pug mill was built. It consisted of a wooden box about four feet square and six feet deep, sitting on ground level. A shaft with push paddles was set in to mix and pulverize the clods and lumps and, at the same time, to push the mud downward. At the bottom on one side was a hole about eight inches square where the mixed mud was forced through, ready for moulding. At the top of the shaft, a long "sweep" was attached, extending out on both ends, one to

BRICK MAKING at HOLLY SPRINGS, MISSISSIPPI

Visitors of the Holly Springs, Mississippi Garden Club are told, "the brick were made by hand right here on this spot over a hundred years ago." Many will be interested to know that these brick are still being made today "right here on this spot."

balance and the other to receive the mule power necessary for propelling.

In making brick, the mud was mixed in a pit on the opposite side from the mud box, by adding water to the clay and dumping a shovelful at a time into the top of the mill. When the mud came out, the moulder standing in a pit received it. He struck off lumps, moulding and sanding it, dumping it with force into the moulds, and striking off the residue. The moulds were then emptied by dumping on wood pads, and the contents moved to covered racks where for days (depending on the weather) they were air dried and then moved to the kiln location for burning. The racks were lined up row by row between the pug mill and kiln location. The unburned brick were then stacked in one large pile, two by two, changing direction every fifth or sixth course, with air space between, in a manner to prevent falling. From fifty to one hundred and fifty thousand formed a "kiln," which resembled a large bee hive. Then the whole kiln was sealed with a two layer brick and clay cover and plastered over with clay, leaving "eyes" or fire boxes extending entirely through near the ground at the bottom. Now the kiln was ready for burning.

Here was the real test of skill. Each new location was a problem for the brick maker because it offered a different clay. The secret of good brick is the burning, and no two clays will take the same heat treatment. Ordinarily, however, a slow fire was made for three or four days to dry out any excess moisture in the brick, then increased to a maximum of 2500 degrees for two weeks, at the end of which time the "eyes" were sealed and the kiln allowed to cool a week before opening. All of the brick were used, hard ones and soft ones; each had a use.

At Holly Springs, Mississippi, there is today a plant making brick exactly as they were made one hundred years ago, mule power and all. The illustration here was sketched at their plant.

Wherever in the South the raw materials were readily available, brick was preferred for building because of its beauty and its safety against fire. Sometimes, indeed, brick was also the cheapest, because where the owner had available clay, raw materials and slaves in abundance, he would have had to purchase lumber at a premium. However, in other communities brick making was very expensive because there was no local clay available. When Williamson Glover built Rosemount in Alabama, records show that he first built of brick but later destroyed the house and rebuilt of frame, because the brick would not hold up. This would seem to substantiate the fact that rich farming soil such as that in the Black Belt of Alabama does not make good brick.

In Louisiana, the true French houses were built of heavy cypress framing and filled in with a mud and moss mortar, then plastered, or boarded on either side, inside and out. Bricks were used for chimneys only, where they were used at all. Such houses were much more comfortable in that locality than brick ones; but in Tennessee they would have cost more than a good house of Tennessee clay brick, admittedly in those days the best building material. So we conclude that the raw materials available determined more than any other factor what kind of houses predominated in any vicinity.

After the brick had been made, the masonry craftsman prepared to build the foundations, inside and outside walls, and chimneys. Most foundations of brick houses were of brick, rock being used in some cases where available.

Ordinary brick foundations and walls were constructed very much as they are today, but in most cases with more care and pride given to the bond, both exposed and hidden. Then, too, there were a great many moulded bricks used for doorways and cornices, such as those found in Fairview. Specially formed "jack arch" brick were used extensively for spanning a square top opening, where we now use a steel lintel. Great care was given to radial brick for building columns, which were in turn stuccoed. In the Little Chapel of the Cross in Mississippi, special butress caps and inserts were burned of clay, or terra cotta as we call it today. Then there were those fine old "pavers"—six, eight, ten, and twelve inches square and three inches thick on porches, terraces, and gardens.

All in all, the art of brick-making was a worthy craft. Into the making of every one of these old brick went skill, intelligence, and quality. The making of good brick required years of training and a natural aptitude for handling materials. It is very certain that specially qualified craftsmen and not ordinary slaves moulded the beautiful brick in the houses of the South.

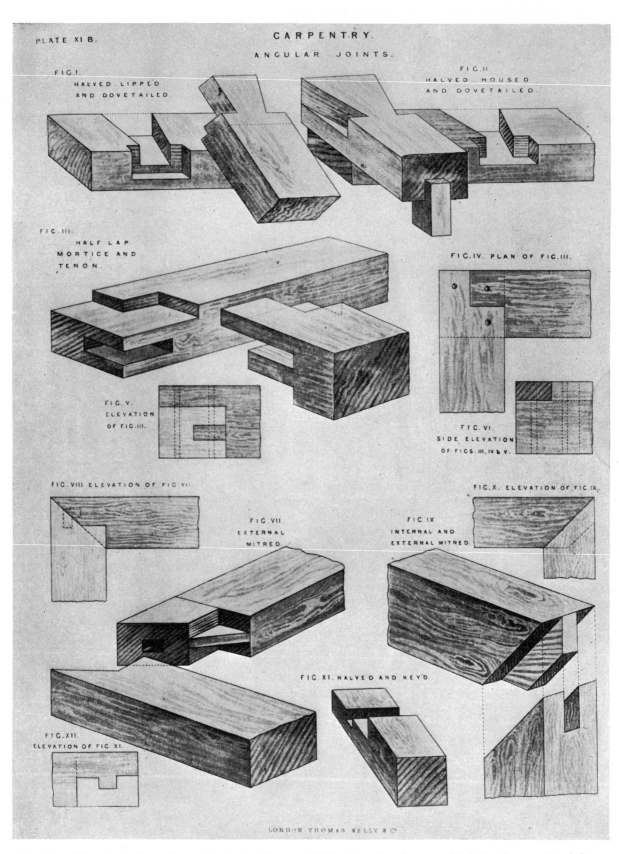

Reprinted here is a plate from a book by Thomas Kelly, showing the art of joining in angular joints. The reader must keep in mind that no nails or screws were used in carpentry in those days.

CARPENTRY

CARPENTRY, joining, stair building, and cabinet making all applied skill to the execution of design in wood. In the earliest days of the lower South, when the French and Spanish Colonial civilizations were struggling for existence, these crafts were all important. Practically every structural and prime part of a house was built of wood. Even the foundations were cypress logs squared and laid flat on the swamp land, and many of these foundations are in existence today. Chimneys were built of wood frames and split lath, lined inside and out with mud. These people retained even beyond the formative period their preference for wood. Many of the first Louisiana brick houses were built of wood frame and filled in between and veneered with brick, as is the case at Ormand Plantation, north of New Orleans on the Mississippi.

The Americans approaching from the east also built entirely of wood at first, as exemplified in their log houses; but as soon as they had time for permanent homes they either refined their wood working or sought more permanent materials, such as brick or stone. Frame construction was as a rule much on the order of early English and French Normandy methods, only heavier in the earlier work. All structural timbers, of course, were hewn, jointed, and pegged, and this part of the work became a specialty among craftsmen who were known as "joiners of framing." Saw mills were scarce and sparsely distributed until the advent of the steam engine; so all rough framing was formed by hewing or adzing the logs square.

Lumber to be worked into planed material for weather-boarding and finish parts was whip-sawn out of the logs. For whip-sawing, the log was laid on a platform parallel to the ground—a stage built above and a pit dug beneath. A long saw blade, six to eight feet long, similar to our crosscut saw with a double grip handle on each end, was pushed and pulled, or "whipped" back and forth through the log. Power was furnished by one slave in the pit below on one end of the saw and a second slave on the platform above on the other end. Fresh men were available for rest shifts so that planks could be peeled off at no slow rate. After rough planks were thus obtained, they were ripped to widths, planed, and moulded by hand tools, similar to our present-day carpenter planes. They were then made ready for all kinds of interior and exterior finish cabinet work.

In restoring early houses, when we replace old work, we find slight variations in mouldings which make it difficult to match them with our machine-made mouldings. There were carpenters who made the sash, frames, panels, chimney pieces, doors, and cabinets, all right there on the job. In the early part of the last century joining was still the method for fastening the pieces together and otherwise assembling the various parts into the whole. Everything was solid material, as veneering was then an unknown art to house builders. Later, of course, with the advent of the planing mill, this arduous labor was diminished, and, as we have discussed in previous paragraphs, fewer experienced craftsmen were necessary to hand-mould every door, sash, and mould.

A modern architect is impressed with the versatility and skill of the early Southern carpenter.

From the many books written for his guidance I have selected a page from an early one showing methods of framing and joining. I reproduce it here to illustrate the precision and care necessary to perform one simple operation in wood work.

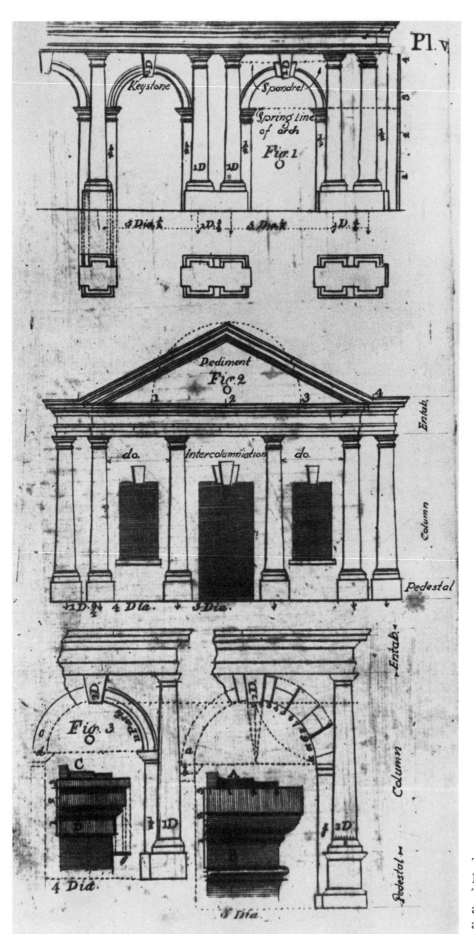

This is Plate No. VI of John Norman's book, showing the Tuscan Order in use with other architectural motifs such as arches, pediments, windows, etc.

Suppose a Southerner about 1835 were thinking of building a house. The floor plan was easy to decide upon; in fact, it was already settled and had been for a generation. The principal concern, the one thing most anxiously discussed, was the portico — porch — veranda — gallery — call it what you will; it had to have white pillars. He pored over the available books on architecture for advice and models. One can turn the pages of a typical guide to see what styles were available. The pages reproduced here from favorite authorities of the day will serve the reader, as for the early Southerner, an illustrated glossary of terms: base, plinth, pedestal, shaft, capital, architrave, etc.

Next came the selection of the "order of architecture." From the Greek he had three choices: Doric, Ionic, Corinthian; and from the Roman, five: Tuscan, Doric, Ionic, Corinthian, and Composite. The illustrations gave him complete information as to the correct proportions, all figured out mechanically. This process has been frowned upon by architects of all times but was very useful to practical builders and laymen. His text-book furnished the classical rules and proportion, but gave him the liberty to modify them to suit his own needs. This pragmatic note in the early architectural books encouraged the Southerner to make his great contribution to domestic architecture.

An instance of this encouragement toward freedom is found in Asher Benjamin's *A New System of Architecture* (1806):

"To retain an exact imitation of these noble products of former times is too expensive for American consumption except on public buildings and this volume proposes to simplify and lighten their heavy parts and thereby lessen expense of labor and materials . . . 1/6 and 1/4 part. At the same time preserve proportions and harmony of parts, etc."

He further proposes to "make alterations in proportions of different orders, by lengthening the shafts of the columns" or diminishing the diameter of them, but "letting the entablature and pedestals bear nearly the same." He also gives a great variety of figures, cornices, and capitals "calculated both for wood and stucco." That a number of people agreed with Benjamin is evidenced by the houses we have discussed in previous chapters.

A page reproduced here from John Norman's *The Town and Country Builder's Assistant* (1786), which shows "the Tuscan Order in use," was no doubt welcomed by many a builder, like Samuel Young, who seemed to have been troubled over the result of his "combination of elements" at Chrétien Point; that is, the spacing of columns and the forming of the entablature at roof line.

It must be remembered that the ideal classic orders were all intended for stone as a material, while in the South with the exception of Kentucky and Tennessee, other material was chiefly used. In Natchez and the Felicianas, as well as in Louisiana, brick and stucco prevailed. In the case of brick and stucco, local tradition developed craftsmen who evidently made a specialty of this art, judging from the fine shafts and bases of the Corinthian order found in Windsor Castle in Mississippi, where the brick were "radial" and the stucco fluted correctly from base to capital, a distance of some thirty-five feet. Where wood was used for columns, as in Mississippi and Alabama, there were steam planing mills making them complete and shipping to all points. Some workmen, however, shaped shafts out of solid logs and carved the bases and caps on the ground; others with a desire for local tradition and not knowing where to purchase, made their shafts octagon-shaped instead of round or square.

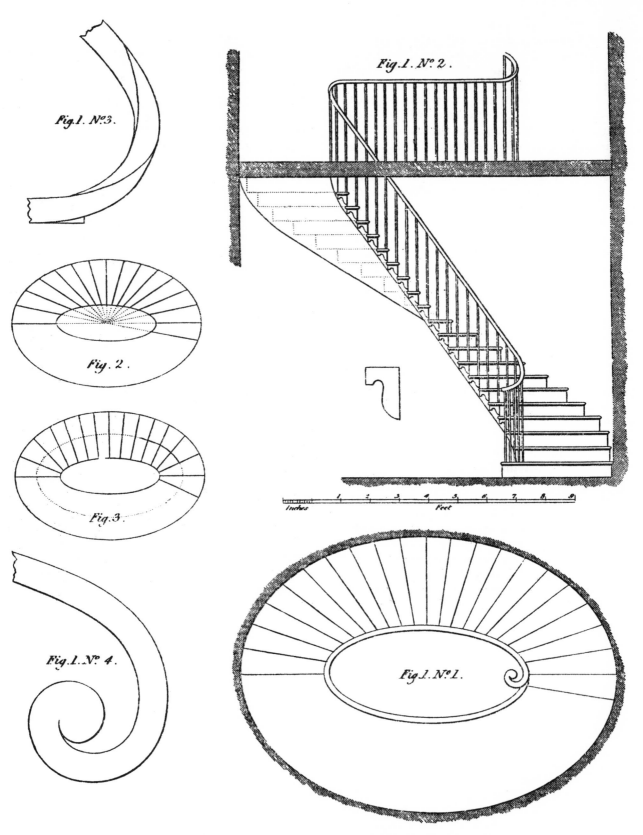

Fig.1. Nº3.

Fig.1. Nº 2.

Fig. 2.

Fig. 3.

Fig.1. Nº 4.

Fig.1. Nº 1.

Inches   Feet

This plan and geometric diagram of an elliptical stair was published by Tho. Kelley—17 Paternoster Row—London, England—Jan. 12, 1848. Said Peter Nicholson, "to execute this design would be the most elegant and useful qualification which a workman can possess."

STAIRCASES

NEXT to his white pillars, the Southerner coveted a circular stair, and rightly so. Was not the stair hall the center of activity and the hub of the plan? Was it not well to approach the second floor with the ease and grace befitting its importance? Was it not desirable to impress the person entering the house with a work of art designed to incite admiration? And he found encouragement in many early books devoted exclusively to stair design and construction: square stairs, dog-leg stairs, circular stairs, free-standing spiral stairs, oval spiral stairs.

As early as 1750, Abraham Swan published *The Builder's Treasury of Stair Cases*. In 1838, Minard Lafever issued *The Modern Practice of Staircase and Handrail Construction*, illustrating his shrewd development of "Descriptive Carpentry." In 1847, Peter Nicholson wrote *A Treatise on the Construction of Stair Cases and Hand Rails*. His preface reads in part:

"The skill and geometrical knowledge required to form the handrail of a stair, so as to range over its plan, and to form an agreeable figure to the spectator in every position, has at all times been considered one of the most elegant and useful qualifications which a workman can possess. It has been justly regarded as the very summit of the joiner's art; and though considerable skill is required in descriptive carpentry, no workmen in any branch of the building art have acquired so much celebrity as those who understand this principle, and which they can apply with facility in the construction of handrails."

From Nicholson we have reprinted Plate LXIII.

Evangeline Country, Louisiana

BAYOU TECHE TODAY

It must have been a privilege to live in a Bayou Country where live oaks formed giant Gothic archways, draped in their silvery Spanish moss. Such deep and spreading silence will inspire more white pillars and remind those of another age of that which we know too well—that "nothing is built by man but makes, in the end, good ruins."

CHAPTER IX

A Pilgrimage

TO

CHAPTER IX

A Pilgrimage

INCREASING interest in pilgrimages to the Far South has brought many requests for information about an itinerary. Where to go, what to see, and when to see it. While this volume illustrates some seventy-five selected houses only, there are hundreds worth seeing, both in communities from which these types were chosen, and in other locations.

Using the early map of the lower valley country, shown in Chapter I, page 14, one will note the concentrated trends of civilization in the year 1830 (dotted area). In choosing an itinerary, these early communities with their various highways and byways should be followed as closely as possible. I have arranged a suggested itinerary in sections more or less in keeping with the Chapters of this volume but not necessarily in chronological order. One may choose any one—or all. It is recommended that Highway (State) maps be marked to agree with the itinerary and used for highway travel, referring to the 1830 map (Chapter I) only for historic data and early atmosphere.
To proceed accordingly:

Lexington, Kentucky, is the center of the Bluegrass Country. With Lexington as a hub, select the pikes and work out in the mornings and back in the afternoons, with side trips to Louisville and Ludlow. The Kentucky Highway Department, Garden Clubs, A.A.A. and local Chambers of Commerce all have available literature on Kentucky routes of interest.

Follow U. S. Highway 31 E. to Nashville and environs. With Nashville as a hub, work in and out to Clarksville, Gallatin and Murfreesboro. The local motor club and State Highway Department can furnish detailed information as to Highways, etc.

To follow the Old Trace Route through to Natchez, one will have to diverge considerably. Keep in mind, however, that the original route was at times necessarily changed from ridge land to ridge land and from ford to ford, according to the ever-changing river beds. Today, there is much argument about the true route of the Old Natchez Trace because the United States Government is building a new super-highway, and one requirement is that it is to follow as closely as possible the original trace.

Modern highway travel will carry us on a course which will cross and re-cross the Old Trace Route many times.

From Nashville, U.S. 31 will parallel the Trace fairly close through Franklin, Tennessee—to Columbia, Tennessee—on to Pulaski, Tennessee. Follow on through to Athens, Alabama, and Florence, Alabama, on U.S. 31 with a side trip to Huntsville, Alabama; or take U.S. 64 to Lawrenceberg, Tennessee, and U.S. 43 to Florence, Alabama, and make side trips to Athens, Alabama, and Huntsville, Alabama. From Florence, Alabama, take U.S. 72 to Corinth, Mississippi, and U.S. 45 to Tupelo, Mississippi, where one may take U.S. 78 to Holly Springs, Mississippi, and Columbus, Mississippi. From Columbus, Mississippi, take U.S. 82 to Greenwood, Mississippi, and inquire locally as to location of Malmaison. Reach U.S. 51 via local road and follow same to Madison, Mississippi, and inquire about Annandale, which is on a local road west of the Highway. Returning to U.S. 51, proceed to Jackson, Mississippi, and then U.S. 80 to Vicksburg, Mississippi. From Vicksburg, U.S. 61 goes to Port Gibson and Natchez.

At Natchez, Mississippi, one will find local Garden Clubs active in giving information and guide service. Their Pilgrimages are held in March of each year and I recommend that visits be made at this time.

Leaving Natchez, follow U.S. 61 to Woodville, Mississippi, and St. Francisville, Louisiana—and State Road 35 to Jackson, Louisiana, and Clinton, Louisiana. Inquire locally at St. Francisville about houses listed in the Feliciana Parishes. This country is easily visited via the Natchez Garden Pilgrimages at the same season.

From St. Francisville, Louisiana, one may ramble along the Old River roads on either side, crossing and re-crossing at will by ferries which are clearly marked all along. (One will more or less have to forget the U.S. Highway system; to see Louisiana, seek the byways.) When at Donaldsonville, Louisiana, follow Bayou Lafourche up to Thibadoux, Louisiana, cross and return on the opposite bank. Continue down to New Orleans and below until the trails play out.

Bayou St. John is in New Orleans. In New Orleans, all hotels are equipped to furnish information and guide service and there are excellent books one can purchase (See Bibliography) written exclusively on the City. The Louisiana countryside is also well covered by local authors.

Bayou Teche and the Cane River Country is best reached by Highway U.S. 90 to Lafayette, Louisiana. This highway closely follows the Bayou, but the old byways of the Teche are still in use and one should follow them whenever and wherever they lead off. Then, too, from the time one crosses Bayou Lafourche, there are continuous numbers of little interesting streams leading off westward that when followed will always prove worthwhile. At New Iberia, Louisiana, follow the River Road to St. Martinsville, Breaux Bridge and on beyond the Teche to Opelousas. Beyond Opelousas, a number of byways reach toward Natchitoches, however, through Alexandria, along the banks of the Red River, will prove an interesting trip.

From Opelousas, Louisiana, one can return eastward along U.S. 190, crossing Bayou Atchagalaya and turning off southwardly on State Road 65 to pass through Moringauin, Louisiana; Rosedale, Louisiana, and Grosse Tête, Louisiana. This is the Bayou Grosse Tête and Bayou Moringauin country.

Continue on State Road No. 65 to Plaquemine, or via State Road No. 1 to Port Allen and Baton Rouge, Louisiana, and in turn take the "Air Line" Highway U.S. 61 to New Orleans.

From New Orleans, take U.S. 90 along the Gulf Coast through Gulfport and Biloxi to Mobile and the Black Belt Country of Alabama. Be sure to stop at Bellingrath Gardens, just before reaching Mobile. (One could spend the day here whether he be a garden enthusiast or not.) In Mobile, the Chamber of Commerce has an excellent map and guides to the City; or local hotels have the usual travel accommodations.

To see the Alabama Black Belt, leave Mobile and again forget about concrete highways. Follow the roads northward, staying as close to the Alabama River as possible until you reach Camden, Alabama. Make a side trip to Greenville, Alabama; then go north to Montgomery and Tuskegee. These two latter towns are geographically out of the Belt, but are in the same economic sphere. Cross the River northwest from Camden and go on to Selma (or if one takes the side trip to Montgomery, head west from Montgomery). Between Selma and Montgomery is the old ghost town of Lowndesboro. Near Selma, in Dallas County, is another antique spot—the town of Summerfield. The graveyard tells its history. From Selma westward to Demopolis, we come to the Tombigbee and Black Warrior River Junction; then northward through Eutaw to Tuscaloosa and U.S. Highway 11.

The itinerary above will take one through all the country covered by this volume. Below, the houses recommended for visiting are listed with location opposite each.

Large communities and cities, such as Nashville, Mobile, Natchez, New Orleans, etc., have so many places of interest, it is impossible to include them all here. Local Chambers of Commerce, garden pilgrimages, organizations and sight-seeing services can furnish detailed information and guides, as well as assist in many other ways. In these cities I include only suggestions of outstanding types of white-pillared houses.

See bibliography for recommended books on local houses.

In the list that follows, many of these houses were visited as far back as 1930 and may have since been destroyed without my knowledge.

CHAPTER II, PART 2

KENTUCKY AND THE BLUEGRASS COUNTRY

(* Houses illustrated herein)

Lexington, Fayette County, Kentucky
*Rose Hill
Bodley House
Ashland
*Mount Brilliant—Russell Cave Pike,
 near Lexington
Mansfield
Gratz Home
Botherum
General Morgan's Home (Hopemont)
Greentree—Highway 68—Maysville Pike
Cave Hill Farm—Harrodsberg Pike

Helms Place—Harrodsberg Pike
*Walnut Hall
Kilmore (Lewis Manor)
Mansfield—Beyond Ashland on Richmond Road
Malvern Hill
Gothic houses
 Ingleside
 Landoun
 Ashland
Frankfort, Franklin County
Liberty Hall
Scotland—On Versailles Road

Woodburn—On old Frankfort Turnpike Road
Danville, Boyle County
    Claverhouse—Perryville Road
    Old Chestnut Home
    Adams Home
Harrodsberg, Mercer County
    Clay Hill
    *Diamond Point
    Maberly House
Bardstown, Nelson County
    Wickland
Paris, Bourbon County
    The Larches

Roccliegan—On Lexington Road
Glen Oak—North of Middleton
The Grange
Louisville, Jefferson County
    Farmington—Old Speed Home
    Spring Station—Lexington Road
Ludlow
    Elmwood Hall
Versailles
    Old Adam Childers House
Georgetown
    Brooker House
    Showalter House

CHAPTER II, PART 1

NASHVILLE AND HER NEIGHBORS

Grundy County, Tennessee
    *Old Beersheba Inn
Lebanon, Tennessee
    Caruthers Home
    Nathan Green House
Clarksville, Tennessee
    Emerald Hill
    Lewis House
(Near) Gallatin, Tennessee
    *Cragfont
    Rock Castle
            (between Gallatin and Hendersonville)
    Spencer's Choice
            (between Gallatin and Hendersonville)
    *Fairview
            (between Gallatin and Hendersonville)
    Foxland Hall
    Rosemont

(Near) Nashville, Tennessee
    *Hermitage
    Cleveland Hall
    Tulip Grove
Hendersonville, Tennessee
    Hazel Path
Nashville, Tennessee
    *Belmont
    *Belle Meade
    Belair
    Mile End
    Glen Leven
    Kingsley
    Travelers Rest
    Riverwood—On Porter Road
    Sunnyside
    Rockeby
    *Tennessee State Capitol Building

CHAPTER III

ALONG THE NATCHEZ TRACE

Franklin, Tennessee
    *Carnton
Springhill, Tennessee
    Cheairs Place
Columbia, Tennessee
    Mays-Hutton Place
    Polk Home
    Beechlawn Hall
    Pillow-Haliday Place
    Hamilton Place
    Mercer Hall
    Wilson Place
    Skipworth Place (6½ miles from
            Columbia-Williamsport Highway)

*Rattle and Snap (near Columbia)
*Clifton Place (near Columbia)
Mount Pleasant, Tennessee
    Manor Hall
Murfreesboro, Tennessee
    Grantlands
    *Marymont
    The Crest
Pulaski, Tennessee
    Crescent View
    Colonial Hall
    Flournoy-Walker House
    Ballentine House
    Clifton Place—Wales—Near Pulaski

Bolivar, Tennessee
  Mecklenburg (Polk Mansion)
  McNeal House
Lauderdale County, Alabama
  *The Forks of Cypress
Limestone County, Alabama
  *Belle Mina
  Francis Snow Pryor House (Athens, Ala.)
Columbus, Mississippi
  *Waverly—Near Columbus
  *Humphrey's Place
  *Old Richard's Home
  *Street House (The Colonnades)
  *Pratt House
  *Elias Fort House
Aberdeen, Mississippi
  Reubin Davis Home

*Sykes Home (The Magnolias)
Corinth, Mississippi
  Curlee Home
Holly Springs, Mississippi
  The Walter Place
  Gray Gables
  Fort Daniel Place
  Featherston Place
Carroll County, Mississippi
  *Malmaison (ask at Greenwood)
(Near) Madison, Mississippi
  *Annandale
  *Chapel of the Cross
  *Englesides
Port Gibson, Mississippi
  *Windsor

CHAPTER IV

NATCHEZ AND THE FELICIANAS

Natchez, Mississippi
  *Concord   Richmond
  *Gloucester   Stanton Hall
  *Arlington   Monteigne
  *Rosalie   Monmouth
  *Auburn   The Briars
  *D'Everaux   Connelly's Tavern
15 miles—Woodville, Mississippi
  Richland
(Near) St. Francisville, Louisiana
           (W. Feliciana Parish)
  *Rosedown

*Greenwood
Oakley
*Afton Villa
Ellerslie
Clinton, Louisiana
  Chase House
  Stone House
  Silliman College
Clinton Vicinity—E. Feliciana Parish
  *Asphodel
6 Miles from Jackson, Louisiana
  The Shades

CHAPTER VII

ALONG THE LOWER MISSISSIPPI RIVER

*Klein House—Vicksburg Mississippi
*Parlange—Near New Roads—across from
    St. Francisville
*Labatut—Near New Roads
*Belle Helene
    or
Ashland—Near Geismar, Louisiana
*Uncle Sam—Near Convent, Louisiana
*Keller Plantation—Hahnville, Louisiana
*Jefferson College—Lutcher, Louisiana
*Oak Alley—Near Vacherie (50 miles north
    of New Orleans)
*Evergreen—Near Donaldsonville
*Ormond—Near New Orleans
*René Beauregard—Near New Orleans

*Three Oaks—Near New Orleans
D'Estrehan—North of New Orleans (east Bank)
Waldheim—Near La Place, Louisiana
Convent of the Sacred Heart—Few miles
    beyond Jefferson Avenue
Colomb—Halfway between Convent & Burnside
Burnside Chatsworth—15 miles south of
    Baton Rouge
Concord—10 miles south of Baton Rouge
Laycock House—Baton Rouge Suburbs
Dougherty-Prescott—North Street—Baton Rouge
West Bank of the Mississippi River
    —below New Roads
Lakeside
Riverlake

West of Baton Rouge and Port Allen
—20 Miles west to Bayou Grosse Tête
and Bayou Moringauin
Old Dickinson—Rosedale
The Mounds—Rosedale
Shady Grove—Between Rosedale and Town of
Moringauin facing Bayou Grosse Tête

Belmont—1 mile west of town Moringauin
on Bayou M.
Llanfair—West Bank—below Port Allen
Belle Grove—Port Allen
Le Breton Home—Westwego, Louisiana
Belle Chase—5 miles back of Gretna
The Keller House—Hahnville, Louisiana

CHAPTER VI

BAYOU TECHE AND BEYOND

*Shadows-on-the-Teche—New Iberia
*Darby Place—New Iberia
Oaklawn—Franklin
Don Caffrey House—Near Franklin
Dixie
Old Inn—St. Martinsville
Payne—5 miles north of Washington
*Chrétien Point—Near Opelousas
(The Gardner House)
Cane River and Natchitoches
Bermuda—Near Natchitoches
*Melrose—Near Natchitoches

Marco—Near Natchitoches
Bayou Lafourche
*Belle Alliance—Near Donaldsonville, La.
*Madewood—Near Napoleonville, La.
*Woodlawn—Near Napoleonville, La.
Bayou St. John
*Schertz House (Spanish Custom House—
New Orleans)
Grandeur—And other houses on Moss Street
facing Bayou
See any travel Bureau in New Orleans for other
houses in and around New Orleans

CHAPTER V

MOBILE AND THE ALABAMA BLACK BELT

Mobile, Alabama
*Judge John Bragg (General Bragg Home)
*Oakleigh
William A. Dawson House
Jonathan Emanuel House
Bishop Portier House
Springhill, Alabama
The Eslava Home
Dawson Perdue Home
Assyrian Grove
Camden (vicinity) Wilcox County, Alabama
McDowell Place
Robert Tait Plantation
Felix Tait Plantation
Frank Tait Plantation
Montgomery, Alabama
Gerald House
Stone-Young House (8 miles from Montgomery)
Pollard Mansion
Housman House
Tuskegee, Alabama
Warner-Alexander House
Selma, Dallas County, Alabama
Parkman-Gilman

Lawndesboro (between Montgomery and Selma
on Alabama River)
Haygood House
Greensboro and Summerfield, Alabama
Magnolia Grove
Moore House
Hudson Summer House
Sturdivant-Hartley House
King House and Graveyard
Demopolis, Alabama
*Rosemount
Gaineswood
Kirksey House (Eutaw—Green County)
Thorn Hill (10 miles from Eutaw)
Tuscaloosa, Alabama
Dearing House
The Governor's Mansion
Eddins House
*Gorgas House
Tyson Home
Reese Home
Turner-Dixon House
Dr. Deal Home
President's Mansion (University of Alabama)

BIBLIOGRAPHY

EARLY TRAVELOGUES, EMIGRANT GUIDES, JOURNALS, ETC.

From Frontier to Plantation in Tennessee
Thomas Perkins Abernethy (1932)
The Blugrass Region of Kentucky
James Lane Allen (1892)
Paddle Wheels and Pistols
Irwin Anthony (1931)
Heart Whispers
William Atson (1859)
Biography of Audubon
S. C. Arthur (1937)
Flush Times in Alabama and Mississippi
James J. Baldwin (1853)
The Englishman in America
J. Benwell (1840)
Alabama in the Fifties
M. C. Boyd (1931)
A Diary from Dixie
Mrs. Mary Boykin (1906)
The Glory Seekers
Wm. H. Brown (1906)
Early Times in Middle Tennessee
John Carr (1857)
A Belle of the Sixties
Mrs. Virginia Clopton (1904)
Scenes in the South
James R. Creecy (1860)
Mystery of Lafitte's Treasures
Dobie
On the Trail of the Pioneers
John T. Faris (1920)
Wayside Glimpses, North and South
Lillian Foster (1859)
Old Times in Tennessee
W. C. Guild (1878)
Journal
Sister Madeline Hachard
The Southwest by a Yankee
Joseph Holt Ingraham (1835)

The Sunny South
Joseph Holt Ingraham (1860)
Ker's Travels: 1808-1816
Henry Ker
Journal of a Residence on a Georgia Plantation
Frances Anne Kemble (1864)
My Southern Friends
Edmond Kirke (1862)
Creole Families of New Orleans
Grace King (1921)
Ten Years on a Georgia Plantation
Leigh (1860)
Evangeline
Henry Wadsworth Longfellow (1849)
A Second Visit to the U. S. A.
Sir Charles Lyell F.R.S. (1849)
The Southern Ladies Book
Alexander B. Meek
Travels
André Michaux (1793)
Travels to the West of the Allegheny Mountains
F. A. Michaux (1805)
The Historic Blue Grass Line
Published by Nashville-Gallatin
Interurban Ry.
Louisiana
Albert Phelps (1905)
Historic Towns of the Southern States
Lyman P. Powell (1900)
Biography of Audubon
C. M. Rourke (1936)
Seven Lamps of Architecture
John Ruskin (1849)
Americans As They Are
Charles Sealsfield (1828)
Life on the Mississppi
Mark Twain (1883—first edition)
Biography of Audubon
A. J. Tyler (1937)

HISTORY

Epic of America
James T. Adams (1933)
History of Alabama
W. Brewer (1872)
Daniel Boone and the Wilderness Road
H. Addington Bruce (1910)
The Old South
R. S. Cotterill (1937)

Tour to the Western Country
Fortescue Cuming (1804)
The Cotton Kingdom
W. E. Dodd (1921)
The Southern Plantation
E. P. Gaines (1924)
Colonial Mobile
P. F. Hamilton (1897)

Historic Highways of America
        A. B. Hulbert (1903)
Tennessee, The Volunteer State
        John Trotwood Moore (1923)
Romantic Passages in Southwestern History
        A. B. Meek (1857)
Life and Labor in the Old South
        U. B. Phillips (1929)
The Winning of the West
        Theodore Roosevelt (1920)

History of Mississippi
        Dunbar Rowland (1925)
A Southern Planter, Thomas Smith Gregory Dabney
        Susan D. Smedes (1890)
View of the Valley of the Mississippi
        H. S. Tanner (1832)
A Topographical Description of the Western
    Territory of North America—1779

TECHNICAL

Architecture
A New System of Architecture
        Asher Benjamin (1806)
The Rudiments of Architecture
        Asher Benjamin (1814)
Beauties of Modern Architecture
        Minard Lafever (1839)
Journal of Latrobe
        Benjamin Henry Latrobe (1796) (1820)
Domestic Architecture of the Early American
    Republic
        Howard Major (1926)

Spanish Colonial Architecture in the U. S. A.
        Rexford Newcomb (1937)
Civil Architecture
        Edward Shaw (1834)
Country Villas
        Sloan
The Antiquities of Athens
        James Stewart & Nicholas Revett (1787)
Treatise on Architecture
        Thomas Young (1846)
The British Architect (American Edition) (1775)
Tuileries Brochure

Early Carpentry and Building Construction

The Country Builders Assistant (1805)
The Practical House Carpenter (1832)
The American Builder's Companion (1816)
        All by Asher Benjamin
A History of Architecture
        Sir Banister Fletcher
The Young Builders General Instructor (1828)
The Modern Builders Guide (1853)
The Modern Practice of Staircase and Handrail
    Construction (1838)
        All by Minard Lafever

The New and Improved Practical Builder (3 vols.)
    (1848)
(Later—An American Edition of the New and
    Improved Practical Builder)
The New Carpenter's Guide (1825)
A Treatise on the Construction of Staircases and
    Handrails (1847)
        All by Peter Nicholson
The Town and Country Builder's Assistant (1786)

Furniture and Decorator Arts

The Encyclopedia of Furniture
        Joseph Aronson
Period Furnishings
        C. R. Clifford

The Practical Book of Interior Decoration
        Harold Donaldson Eberlein, et al.
Manual of the Furniture Arts and Crafts
        Johnson and Sironen

Gardens (Landscape Architecture)

Old Time Gardens
        Alice Morse Earle
Gardens
        J. C. H. Forestier

Gardens of Colony and State
        Garden Club of America
Landscape Design
        Henry V. Hubbard and T. Kimball

Encyclopedia of Gardening
  J. F. Louden (1835)
Farm Fences and Gates
  Walter D. Popham

Garden and Design
  Shepherd and Jellicoe
Domestic Manners of the Americans—Early
  Nineteenth Century

SOME CURRENT BOOKS ON THE SUBJECT

Historic Homes of Alabama and Their Traditions
  Alabama Members of National League
    fo American Pen Women
Land Pirates of the Natchez Trace
  Robert M. Coates
In Old Natchez
  Catherine Van Court
New Orleans
  Nathaniel C. Curtis
History of Homes and Gardens of Tennessee
  Garden Study Club of Nashville
Chieftain Greenwood Leflore
  Florence Rebecca Ray

Old Louisiana
Old New Orleans
  Lyle Saxon
Bluegrass Houses and Their Traditions
  Elizabeth M. Simpson
Old Plantation Houses in Louisiana
  William Spratling and Natalie Scott
Old Kentucky Homes and Gardens
  Elizabeth Patterson Thomas
Feliciana
  Stark Young
Creole Families in New Orleans
  Grace King

ARCHIVES AND TECHNICAL LIBRARIES

Library of Congress   Washington, D. C.
Tennessee State Archives   Nashville, Tennessee
Howard Library   New Orleans, Louisiana
Missouri Historic Society   St. Louis, Missouri
Goodwyn Institute Library   Memphis, Tennessee

Cossitt Library   Memphis, Tennessee
Mississippi State Archives   Jackson, Mississippi
Mississippi State College
  Library   Starksville, Mississippi

GLOSSARY

A

Acanthus—A plant whose leaves, conventionally treated, form the lower portions of the Corinthian capital.

Acroteria—The small pedestals placed on the extremities and apex of a pediment.

Architrave—That part of an entablature which rests upon the capital of a column and is beneath the frieze.

Apron—A plain or moulded piece of finish below the sill of a window.

Atazère—A large, elaborately carved whatnot stand.

B

Belvedere—A turret, raised above the roof of an observatory for the purpose of enjoying a fine prospect.

Base of a Column—That part which is between the shaft and the pedestal or, if there be no pedestal, between the shaft and the plinth.

Back-band—A moulded member, notched to overlap another to give additional richness.

Baluster—A small pillar or column, supporting a rail.

Bay—Of columns, the space between two columns.

C

Calaboza—Jail (Sp.).

Capital—The upper part of a column, pilaster, pier, etc.

Chairrail—A moulding placed at the height of a chair, to protect the wall from damage.

Coffered Ceiling—A ceiling divided into panels.

Colonnade—A row of columns.

Colonette—Small slender columns.

Cornice—The projection at the top of a wall finished by a blocking-course.

Cupola—A small room either circular or polygonal standing on the top of a dome.

Cartouche—An ornament, raised, and usually elliptical or oval.

Columniation—The arrangement or system of columns.

D

Dentil—The cogged or toothed member, common in the bed mould of a Corinthian entablature. Each cog or tooth is called a dentil.

Detail—A part of a building, structural or ornamental, small in proportion to the whole.

Dormer—Windows occurring in a sloping roof.

Dovetailing—In carpentry and joinery, the method of fastening boards or other timbers together, by letting one piece into another in the form of the expanded tail of a dove.

E

Entablature—The assemblage of upper parts supported by the columns. It consists of three parts: the architrave, frieze, and cornice.

Entourage—The surroundings of a building; planting, trees, terraces, etc.

Engaged Column—Apparently attached or embedded in a wall.

Escutcheon—The field or ground on which a coat-of-arms is represented.

Elevation—A vertical right line drawing, especially of the sides of a building, hence — the faces of a building.

Étagère—A form of whatnot—with shelves supported by columns.

F

Façade—The face or elevation of a building.

Fluting—Concave channels in columns.

Finial—The flower, or bunch of flowers, with which a spire, pinnacle, canopy, etc., generally terminates.

Fenestration—Arrangement of windows, especially in an aesthetic sense.

Frieze—That portion of an entablature between the cornice above and the architrave below.

G

Gable—A roof which is not hipped or returned upon itself at the ends; the ends of a gable being stopped by carrying up the walls in triangular form of the roof itself.

Gallery—Synonymous here with verandah.

Garçonnière—An outbuilding used as the son's quarters on a plantation; also for unimportant, or overflow guests.

Girandole—A candlestick, usually for several candles; generally elaborately decorated.

Greek Cross—A cross having four arms of equal length.

H

Hip Roof—A roof which rises by equally inclined planes from all four sides of the building.

J

Jack Rafter—A short rafter used especially in hip roofs.

Jack-arch—An arch with a slightly curved or flat soffit, but radiating members—a flat arch.

Jalousies—A blind, shutter, or fixed device, of horizontal slats to admit air while excluding sunlight.

Jamb—The sidepost or lining of a doorway or other aperture.

Joiner—One who does finish or ornamental carpentry.

L

Lintel—The supporting beam over an opening.
Loggia—Similar to verandah, but open generally on one side only—recessed.

M

Modillion—The enriched block or horizontal bracket generally found under the cornice of the Corinthian entablature.
Meeting Rail—That member of a double-hung window, at the top of the lower sash and at the bottom of the upper sash, at which the sash meet when closed.
Muntin—A small member, usually of wood, which divides the lights of windows. These members, when larger, are called mullions.

N

Newel or Newel Post—The post placed at the first, or lowest step, to receive or start the hand-rail upon.

O

Order—The columns with their entablature.

P

Portico—An entrance or porch with a roof supported, at least on one side, by columns.
Parapet—A dwarf wall along the edge of a roof.
Pedestal—The square support of a column, statue, etc., and the base or lower part of an order of a column.
Pediment—A low triangular crowning, ornamented; in front of a building, and over doors and windows.
Perron—A principal flight of steps, with platform and parapet wall.
Pigeonnier—Dovecote, pigeon house.
Pilasters—Flat square columns, attached to a wall, behind a column, or along the side of a building, and projecting from the wall about a fourth or a sixth of their breadth.
Pitch of a Roof—The angle at which the roof goes up.
Plinth—The square block at the base of a column or pedestal.
Punkah—A type of fan hung from the ceiling and moved by a slave by means of a cord and pulleys.
Pavillions—A portion of a building; accented by position, decoration or height.
Parti—The general scheme, fundamental arrangement.

O

Ormolu—A metal composition resembling gold.

Q

Quoins—Large squared stones, or other material, at the corners of buildings.

R

Reveal—The two vertical sides of an aperture.
Ridge-pole—The highest horizontal timber in a roof, extending from top to top of the several pairs of rafters of the trusses, for supporting the heads of the jack rafters.
Riser—The vertical board under the tread in stairs.

S

Saddle-notched—Square cut.
Shaft—That part of a column between the cap and the base.
Shakes—Hand rived boards used as shingles on early houses.
Sill—The piece of timber or stone at the bottom of doors or windows.
Spandrel—The space between any arch or curved brace and the beams, etc., over the same.
Studs—Vertical supporting members in a frame wall.
Stylobate—In classic architecture, a continuous base or substructure on which a colonnade is placed.
Stringer—The exposed, inclined member of a stair into which the steps are worked. It may be open, in which case the steps are exposed on the outside; or enclosed, in which case the steps are hidden from the outside. The stringer is often highly ornamented, especially when of the open type.

T

Triglyph—The vertically channeled tablets of the Doric frieze.
Tread—The horizontal part of a step or stair.
Tracery—Ornamental division and filling in windows, panels, etc.

V

Verde Antique—A dark mottled green marble.
Vignola—Italian architect (1507-73)—Established proportions of Classic orders.